Author: Dr Tony Ayling
Principal photographers: Gary Bell, Rudie Kuiter & Steve Parish

Steve Parish™
KIDS

# Introduction

Sharks and rays are a fascinating and very successful group of fishes. They *evolved* about 400 million years ago and are so well suited to their individual environments, they have not needed to change much since then. They have a reputation as fearsome predators but this is not something they really deserve. Sharks and rays are all *carnivorous*, which means they eat other animals rather than plants, but of all the shark *species* only a few are considered dangerous to people. However, all sharks and rays should be treated with respect and care.

Elephantfish have a strange nose, like a trunk.

The broadnose shark has seven gill slits.

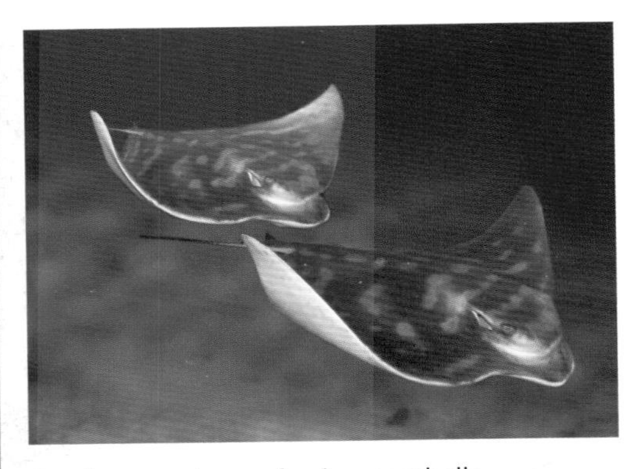

Southern eagle rays feed on seashells.

## WHAT MAKES A SHARK A SHARK?

Most fishes have a hard bony skeleton like humans do, but sharks and rays have a soft, flexible skeleton made from a type of tough, gristly tissue called *cartilage*. However, shark teeth are hard and can be rapidly replaced if they are broken or fall out.

Unlike bony fishes, sharks usually grow slowly and produce only a few young. They *reproduce* using "internal fertilisation", which means that the male's sperm enters the egg inside the female's womb. Male sharks have two *claspers* at the base of the tail that put sperm inside the female's womb, or *uterus*. Unlike mammals, female sharks and rays have two uteruses. Some sharks and rays lay a few large egg cases, but most of them give birth to live young that look like miniature adults.

Sharks have extremely well-developed senses. They have good eyesight, hearing and smell. They also have special "electrosensing" organs on the snout, called ampullae of Lorenzini, which they use to find prey by detecting weak electrical fields put out by living animals.

Tiger sharks grow to 6 metres long and have dark stripes over their backs.

A stingray's underside, showing its gill slits, mouth and nostrils.

## SHARK AND RAY SPECIES

There are more than 300 shark and ray species *inhabiting* Australian waters and they come in every size, from the tiny pygmy shark (which is only about 20 centimetres long when fully grown), to the huge whale shark (which can grow to 12 metres long and is the largest fish in the world). Sharks have many different fins. To identify the species of a shark, scientists look closely at the tail and fins to see their shape, size and position on the shark's body.

## WHAT MAKES A RAY A RAY?

Rays have very large, flattened *pectoral fins* that have merged with the body to form an almost circular disc. The tail is often thin and short, with a large spine on the top near the base. Most of the other fins are small or absent. Rays have their eyes on top of the disc but the nose, mouth and gill slits are underneath. Over time, some rays have evolved to look more like sharks but they still have gill slits underneath the body.

## FACTS ON THE FACT FILE RANGE

Forty-nine species of shark and ray are covered in this fact file, which tells you what each species looks like, what it eats, how it *breeds* and cares for young, whether it is dangerous to humans and its conservation status. Words throughout the book that are in italics are included in the glossary on page 46 to help you increase your knowledge.

# Contents

# White shark *Carcharodon carcharias*

A white shark's sharp teeth are perfect for tearing large prey into bite-sized pieces.

A white shark's dorsal fin often breaks the surface.

The white shark, great white or white pointer (as it is sometimes called), is the most ferocious of all sharks. Most people are afraid of this shark, but it is only really doing what it does best — being one of the ocean's best *predators*! White sharks have a special way of circulating blood through their bodies, which allows them to keep their body temperature warmer than the surrounding water. This gives them more energy than cold-blooded sharks, allowing them to swim faster and further on the same amount of food. They swim with a slightly jerky action.

**WHAT DO THEY LOOK LIKE?** White sharks are blue-grey to grey-brown on the back and white beneath, with a sharp break down the side between the grey and the white. They have heavy bodies, pointed snouts and crescent-shaped tails, five very long gill slits and huge black eyes.

**DORSAL FINS:** White sharks have a high, pointed first *dorsal fin* that is curved at the tip, and a small second dorsal fin back toward the tail. The dorsal fin, along with the tip of the upper tail lobe, can often be seen when the shark swims near the surface. Unlike most sharks, their upper tail lobe is almost the same length as their lower lobe.

**SIZE:** For many years, the maximum size of a white shark was thought to be 36.5 feet (11.5 metres). It was later discovered that this length had been reported incorrectly and was only 36.5 inches (less than 1 metre) long! Some people say these sharks can grow to 8 metres, but the longest white shark measured so far was about 6 metres long and weighed about 3000 kilograms. A 5-metre-long white shark may be between 15 and 25 years old. These large white sharks can live for 50–60 years. A close relative of the white shark was *Carcharodon megalodon*. It became *extinct* millions of years ago but it did reach a length of around 12 metres and must have been a truly terrifying predator.

**WHERE DO THEY LIVE?** White sharks are found from North West Cape in Western Australia around to southern Queensland, but they are usually seen off southern Australia. They are especially common around seal and sea-lion colonies. These sharks are also found off New Zealand, South Africa and California and may occasionally be found in all temperate and subtropical seas throughout the world. White sharks are most often seen in shallow coastal waters but they may also be found in the open ocean far from shore, and sometimes swim down to over 1000 metres.

*Above and opposite (bottom left):* The white shark is grey on top with a white belly and large gill slits.

**TEETH & FEEDING HABITS:** White sharks have widely-spaced, large, triangular teeth in both jaws. Young white sharks feed mainly on fishes and small sharks and rays. Adults may eat marine mammals, including seals, sea-lions and dolphins, although some adult sharks live only on fish. They are active hunters and often attack from below, leaping from the water with their prey in their mouths.

**BREEDING & CARING FOR YOUNG:** We do not know very much about how white sharks breed. No one has ever seen them *mate* and very few pregnant females have been caught and observed. We do know that males reach adulthood at about 8 years old and around 3.7 metres long. Females are larger, reaching adulthood when they are between 12–17 years old and measure 4.5–5 metres. Females give birth to 2–17 young sharks, or pups, that grow inside her uterus by eating other eggs produced throughout her pregnancy. This is called *intra-uterine cannibalism*. After 18 months, the pups are born and measure about 1.3 metres in length.

The white shark is the most feared of all sharks.

*Above:* The nose of a white shark showing the ampullae of Lorenzini (in front of the eye), used for sensing electrical fields put out by prey. *Right*: Ampullae of Lorenzini are electrical sensing pits on the head, a shark's sixth sense.

**MIGRATION & BEHAVIOUR:** White sharks usually travel alone, but may sometimes gather at feeding sites when conditions are right. Satellite tagging programs in Australia, South Africa and California have helped scientists track white sharks to reveal more about their habits.

They don't appear to be territorial, but wander the oceans, making long journeys between feeding or breeding sites, where they may stay for a few days or a few months before setting off again. White sharks have been found to make 5000-kilometre journeys around the Australian coast and often swim between Australia and New Zealand. They sometimes visit a preferred feeding site at the same time each year.

When travelling, they cruise steadily at about 3 kilometres an hour for hundreds or even thousands of kilometres. Research shows that males travel further than females. There is *DNA* evidence that males regularly travel between South Africa and Australia, a distance of over 12,000 kilometres!

**DANGER TO HUMANS:** White sharks are large carnivorous predators and regularly attack large warm-blooded prey, so they are a threat to humans that venture into their environment. White sharks are probably responsible for more fatal attacks than any other shark species. However, many scientists believe that white sharks attack humans by mistake, thinking they are seals or turtles.

**WHAT IS THEIR STATUS?** White sharks are rare and a female only gives birth to about 80 pups over her lifetime, so their numbers are not growing fast. White sharks are now protected in Australia and in many other countries of the world.

*Top and above:* White sharks often launch spectacular surprise attacks on their prey.

*Above and below:* The white shark has *serrated* razor-sharp teeth that are widely spaced in its large jaws.

# Blue shark *Prionace glauca*

The blue shark is an oceanic wanderer and lives in the open sea. It has blue colouring on its back and a white belly, resulting in a "countershaded" colour pattern. This means the parts of the shark mostly exposed to the sun are dark and the parts usually in the shade are lighter. This helps the shark blend in with its surroundings, making it hard for predators or prey to see it from above or below. The blue shark is a large, slender-bodied whaler shark with a sharp, pointed snout and long, swept-back pectoral fins.

A blue shark swims with a pilot fish.

**WHAT DO THEY LOOK LIKE?** As its name suggests, the blue shark has a bright blue back. Its belly is white.

**DORSAL FINS:** The first dorsal fin is relatively low and the second dorsal is about a third of the size of the first. The upper tail fin is much longer than the lower and has a distinct notch. These sharks are graceful and swim in a smooth, curvy path, rather than in a straight line.

**SIZE:** Blue sharks become adults when they reach a length of about 2.2 metres. They can grow to a maximum size of around 3.8 metres.

**WHERE DO THEY LIVE?** Blue sharks are common in most of the world's oceans. They live all around Australia except in the shallow waters of the Arafura Sea, Gulf of Carpentaria and Torres Strait. Blue sharks are abundant off southern Australia. They are found in the open sea and rarely come near the coast. They like a water temperature between 12–20 degrees Celsius and move into deeper, cooler water in the tropics where they may be found as deep as 350 metres.

**TEETH & FEEDING HABITS:** The blue shark has sharp, dagger-like upper teeth for gripping prey and broader, slanted lower teeth for cutting. It feeds mainly on small, open-ocean fishes and squids, gathering in large numbers to feed on squid *spawning* in massive groups offshore.

**BREEDING & CARING FOR YOUNG:** Blue sharks grow fast and become adults in 4–6 years. Courtship

Hunter becomes hunted — a fur-seal attacks a blue shark. Small blue sharks are eaten by larger sharks and even large seals.

The blue shark's white belly is the colour of filtered sunlight.

is vigorous, with the males giving the females so many love bites that the females' skin has to be twice as thick as the males to cope with the damage. Blue sharks have more young than any other shark, with an average of about 40 pups and a maximum of 135. The young sharks are fed through a *placenta* while in their mother's womb, in the same way as human babies. They are 40–50 centimetres long at birth.

From above, the blue shark blends in with the colour of the water.

**DANGER TO HUMANS:** These sharks usually feed on small prey and live far from the coast, so they are not very dangerous to humans in spite of their large size. They may come up close to divers underwater.

**WHAT IS THEIR STATUS?** Common, rarely seen close to land or in tropical waters.

Like most sharks, the blue shark has excellent eyesight.

# Bull shark *Carcharhinus leucas*

A lone bull shark patrolling a deep reef.

Although it is not as well known as the white shark, the bull shark is one of the most dangerous man-eating sharks. Bull sharks live mainly near the coast where most people swim and play in the water, so they are often around humans. Bull sharks are sometimes called river whalers because they can live for a long time, and even breed, in freshwater.

**WHAT DO THEY LOOK LIKE?** This is a heavy-bodied shark with a short, blunt snout and a long, notched tail fin. It has a dark, grey-brown or bronze-coloured back and a paler belly.

**DORSAL FINS:** The first dorsal fin has a sharp, pointed tip and a vertical back edge. The second dorsal is only a third of the size of the first dorsal fin.

**SIZE:** These sharks are fast-growing and become adults at about 6 years and a length of about 2.5 metres. They live for about 15 years and reach a maximum length of around 3.5 metres.

**WHERE DO THEY LIVE?** Bull sharks are found throughout the world in tropical and warm temperate waters. They live all around northern Australia from Perth to Sydney. These sharks are most common in murky coastal waters, especially harbours, large rivers and canals but they may be seen around clear offshore coral reefs.

**TEETH & FEEDING HABITS:** Bull sharks have wide and triangular lower teeth and slightly more pointed teeth in the upper jaw. These teeth are good for cutting as well as gripping small, struggling prey. Bull sharks eat lots of different fishes, large or small, but as adults these sharks like to eat other sharks and rays.

**BREEDING & CARING FOR YOUNG:** Like all whaler sharks, young bull sharks are fed firstly through a *yolk sac* then from a placenta in the mother's womb. Litters range from 1–13 pups. The pups are born after a pregnancy of 10–11 months and measure between 60–80 centimetres at birth.

**DANGER TO HUMANS:** Bull sharks are one of the most dangerous species of shark. Seeing a large and aggressive bull shark when diving can be a frightening experience.

Although they are generally ranked behind the white and tiger shark in the danger stakes, they are more likely to visit places where people swim, so they are often seen as a more dangerous species.

**WHAT IS THEIR STATUS?** Common.

# Tiger shark *Galeocerdo cuvier*

Tiger sharks are large, dangerous predators that have been responsible for many attacks on humans. There is no mistaking a tiger shark. The dark tiger stripes on the back and sides can be seen even on large individuals. They also have a wide head and a huge belly.

**WHAT DO THEY LOOK LIKE?** Tiger sharks have a distinctive blunt head and large black eyes set forward on the snout. The back is grey with darker grey tiger stripes. The belly is pale. The upper tail lobe is very long and thin and does not have an obvious notch.

**DORSAL FINS:** The first dorsal fin of a tiger is relatively low for such a large shark and is usually longer than it is high. The second dorsal is small and set well back near the tail.

**SIZE:** Tiger sharks are usually between 3–5 metres long. The largest measured was around 6 metres. Seven-metre tigers have been seen and bite marks on whale carcasses suggest they may grow even longer.

**WHERE DO THEY LIVE?** Tiger sharks are found throughout the world's oceans in tropical and warm-temperate seas. They are common around northern Australia from Perth to Sydney. They live in all habitats from estuaries and harbours to deep offshore waters but rarely venture below 150 metres in depth.

**TEETH & FEEDING HABITS:** Tiger sharks have hooked teeth that are the same in both upper and lower jaws. The teeth are designed for cutting up large prey animals, rather than gripping small struggling fishes. Tigers have a reputation as swimming "garbage bins". They eat anything and many inedible items have been found in their stomachs. They feed on fishes, other sharks and rays, marine turtles (biting easily through thick shells), octopuses and squids, sea snakes, crayfish and even sea birds. Tigers are more active and aggressive at night but they also feed during the day.

**BREEDING & CARING FOR YOUNG:** Male tiger sharks become adults when they are about 3 metres long. Females are larger. About 12 months after mating they give birth to 10–80 pups that are between 50–75 centimetres long.

**DANGER TO HUMANS:** This is one of the four most dangerous sharks in the world, along with the white, bull and oceanic whitetip sharks. Tiger sharks come into harbours and shallow water where they are more likely to encounter swimmers and divers and have been responsible for a number of deaths over the years.

**WHAT IS THEIR STATUS?** Shark netting has made tiger sharks rare in populated coastal areas but they are still common in offshore habitats.

Tiger sharks are massive predators that hunt in harbours and around shallow reefs — places where they sometimes encounter humans.

# Port Jackson shark *Heterodontus portusjacksoni*

The Port Jackson shark has distinctive harness-like markings.

*Inset right:* A hard, spiral egg case protects the eggs of a Port Jackson shark.

A sharp, poisonous spine sits just in front of the dorsal fin.

The Port Jackson shark has a big blunt head with bony ridges over the eyes and a large notched tail fin. The other fins are also large and there are strong, sharp spines in front of each of the high, triangular dorsal fins. Port Jackson sharks are slow swimmers and these poisonous spines help protect them from larger sharks that try to eat them.

**WHAT DO THEY LOOK LIKE?** These sharks are grey to light brown (or whitish) with a distinct harness-like pattern of darker bars on the body and fins.

**SIZE:** Port Jackson sharks reach a maximum length of 1.65 metres but are generally smaller than 1.35 metres.

**WHERE DO THEY LIVE?** These sharks are found all around southern Australia between Carnarvon in Western Australia to Byron Bay in New South Wales. Young sharks are found close inshore in *estuaries* and bays, but adults move out into deeper water on the outer *continental shelf*. Adults live around rocky reefs and over open sandy bottoms down to depths of 200 metres. Males and females usually live apart outside the breeding season.

**TEETH & FEEDING HABITS:** These sharks have different groups of teeth. Those in the front of the jaw are small and sharply pointed for gripping prey, while those in the back of the jaw are flat and plate-like for crushing food. Port Jackson sharks are *nocturnal*, feeding at night on a variety of bottom living animals. They like to eat "echinoderms", such as sea urchins and sea stars.

**BREEDING & CARING FOR YOUNG:** Mating occurs in winter. The female lays her eggs in rocky reef crevices in water less than 5 metres deep. Each of the 10–15 eggs are about 15 centimetres long with a spiral *flange* around the egg case (*see photo inset above*) to hold it firmly in place until the young hatch, which takes one year.

**DANGER TO HUMANS:** These sharks are harmless but care should be taken in handling captured sharks to avoid being spiked by their poisonous fin spines.

# Crested hornshark *Heterodontus galeatus*

The crested hornshark is a close relative of the Port Jackson shark; however, it has high, bony crests above each eye, higher, more rounded dorsal fins and a different pattern of dark stripes on the side. The dorsal fins have similar protective spines but these are not as long as those of the Port Jackson shark. The skin of the crested hornshark is very rough and feels like coarse sandpaper.

**WHAT DO THEY LOOK LIKE?** Crested hornsharks are light brown with darker bands through the eye and mid body, and dark saddles behind each dorsal fin.

**SIZE:** Most adults are less than 1.2 metres long, but some sharks may grow to 1.5 metres.

The crested hornshark has large bony crests above each eye.

**WHERE DO THEY LIVE?** Crested hornsharks only live on the east coast of Australia, off the southern Queensland and New South Wales coasts between Moreton Bay and Batemans Bay. These strange-looking sharks are usually seen around rocky reefs in depths of more than 20 metres.

**TEETH & FEEDING HABITS:** The front teeth of these sharks are small, with five sharp points each, and are used for gripping prey animals. Those in the back of the jaw are flattened crushing plates.

Crested hornsharks feed on bottom-living *invertebrates* such as sea urchins, shellfish, crabs and shrimps. They sometimes manage to catch small fishes.

**BREEDING & CARING FOR YOUNG:** Females lay large eggs during the winter. Like those of the Port Jackson shark, the eggs are spirally flanged. The eggs also have long sticky hairs rising from the tips. These hairs help attach the eggs to the ocean floor among seaweed and sponges in depths of 20–30 metres.

The eggs hatch after 8 months when the young are about 20 centimetres long. Young crested hornsharks take up to 12 years to grow to adult size. These sharks may make regular *migrations* to and from breeding and feeding grounds.

**DANGER TO HUMANS:** These sharks are harmless to humans, but care should be taken if handling crested hornsharks to avoid the sharp, poisonous fin spines.

**WHAT IS THEIR STATUS?** Crested hornsharks are rare but they may be more common in deeper water and are known to be found to depths of about 100 metres.

Crested hornsharks can grow up to 1.5 metres long.

# Oceanic whitetip shark *Carcharhinus longimanus*

The oceanic whitetip shark has unusually rounded fins.

Oceanic whitetips are one of the four most dangerous sharks, along with the white, tiger and bull sharks. These sharks live in the open ocean far from the coast. They are a large, heavy-bodied whaler shark with very distinctive and large rounded pectoral fins and first dorsal fins.

**WHAT DO THEY LOOK LIKE?** Oceanic whitetips get their name from the mottled white tips on the first dorsal, pectoral, *pelvic* and tail fins. Young sharks less than about 1.3 metres long do not have these white fin tips. There are black markings on the other smaller fins. The back is usually bronze-grey in colour and the belly is paler.

**DORSAL FINS:** The first dorsal fin is unusually wide and high with a broad rounded tip, but the second is small and triangular and set back near the tail.

**SIZE:** Most oceanic whitetip sharks reach a maximum length of about 3 metres, but there is a reliable record of a 3.5-metre shark.

**WHERE DO THEY LIVE?** Oceanic whitetips are found worldwide in tropical and warm temperate seas. They live in offshore waters around northern Australia from Perth to Sydney, but are not found in shallow areas of the Arafura Sea, Gulf of Carpentaria and Torres Strait. These sharks live far offshore in the open ocean and, like most humans, prefer water warmer than 20 degrees Celsius. Oceanic whitetips usually prefer sunlit surface waters but may be found as deep as 150 metres. Oceanic whitetips usually swim slowly near the surface and are equally active both day and night.

These sharks are an impressive and beautiful sight when encountered in their own environment but because of their open ocean habitat little detail is known of their lifestyle.

**TEETH & FEEDING HABITS:** The upper teeth are large and broadly triangular for cutting up prey but the lower teeth are more dagger-like for gripping struggling prey. Oceanic whitetip sharks feed on a variety of open ocean fishes and squids.

**BREEDING & CARING FOR YOUNG:** These sharks become adults at about 1.8–2 metres. After mating, the females develop 1–15 young during a 12-month pregnancy. The young are fed from a yolk sac when small and then a placenta until they are born at a length of about 60 centimetres.

**DANGER TO HUMANS:** Oceanic whitetips are considered to be one of the four truly dangerous sharks. They are very aggressive but are not often encountered by humans because they live so far offshore. These sharks have made many open ocean attacks on humans after air and sea disasters.

**WHAT IS THEIR STATUS?** Very common, although most people never get to see an oceanic whitetip shark because they live so far from land.

## Grey reef shark *Carcharhinus amblyrhynchos*

This shark is also known as the black-vee whaler.

Most people who dive on the Great Barrier Reef have been frightened by grey reef sharks at some time. The main enemies of grey reef sharks are larger sharks like tiger and bull sharks. When they feel threatened by a large shark they will put on an "arched-back" display to warn that they are very fast, agile swimmers and will give a bite if the bigger shark does not back off.

**WHAT DO THEY LOOK LIKE?** A medium-sized whaler shark with a prominent black margin to the tail fin. It has a bronze or grey back and pale belly.

**DORSAL FINS:** The first dorsal is medium-sized with a rounded tip. The second is small and black-tipped.

**SIZE:** They occasionally reach a length of 2.5 metres but most are less than 2 metres long.

**WHERE DO THEY LIVE?** Grey reef sharks are found throughout the tropical Pacific and Indian Oceans. They are common around coral reefs, near drop-offs and lagoon passes, in northern Australia from Carnarvon to Bundaberg.

**TEETH & FEEDING HABITS:** Grey reef sharks have sharply pointed but small teeth. They mainly eat small reef fishes and occasionally squids or crayfish.

**BREEDING & CARING FOR YOUNG:** Grey reef sharks become adults at about 10 years of age. Females give birth to 3 or 4 pups every two years and only give birth to about 15 pups in their lifetime.

**DANGER TO HUMANS:** These sharks make threat displays to divers and may give a quick warning bite. They are not usually thought to be a threat to humans.

## Whitetip reef shark *Triaenodon obesus*

The first dorsal fin and upper tail fin is white-tipped.

Divers and snorkellers on the Great Barrier Reef are very familiar with whitetip reef sharks and see them almost every time they dive on a coral reef. This common species is a slender whaler shark with a small blunt head and distinct white tips on the first dorsal fin and upper tail fin. The first dorsal fin is relatively high and pointed. The second is about two-thirds the height of the first.

**WHAT DO THEY LOOK LIKE?** Whitetip reef sharks are grey on the back and pale beneath.

**SIZE:** They reach a maximum length of 1.7 metres.

**WHERE DO THEY LIVE?** Whitetips are found throughout the tropical Pacific and Indian Oceans. They live all around northern Australia from North West Cape to Gladstone at depths down to 50 metres. They are usually found swimming near the ocean floor and often rest there, or shelter in caves, during the day.

**TEETH & FEEDING HABITS:** Whitetip reef sharks hunt more actively at night and have small, sharp teeth. They feed on small reef fishes.

**BREEDING & CARING FOR YOUNG:** Whitetip reef sharks reach their adult size at 8 years of age, and females give birth to only two pups every second year. They live for about 20 years and each female has only 12 young during her life.

**DANGER TO HUMANS:** Whitetips are not seen as dangerous to humans. However, they may bite if handled or annoyed.

**WHAT IS THEIR STATUS?** Common.

## Australian angelshark *Squatina australis*

Australian angelsharks are also called monkfish.

The Australian angelshark looks like a cross between a shark and a ray. The ray-like head, body and pectoral fins are flattened but the gill slits are on the side of the body like those of a shark. The hind body and tail are similar to that of a slender shark, with two small equal-sized dorsal fins. Two large, fringed sensory feelers or barbels are under the chin.

**WHAT DO THEY LOOK LIKE?** The top of the body is dull grey-brown with lots of small, irregular white spots, and white fin edges.

**SIZE:** These sharks reach a length of 1.5 metres.

**WHERE DO THEY LIVE?** Australian angelsharks are only found around southern Australia between Perth and Sydney. They are common in shallow water through Bass Strait and off south-eastern surf beaches. They are sometimes found in water 130 metres deep.

**TEETH & FEEDING HABITS:** Australian angelsharks have many small sharp teeth in their strong jaws and feed on small fishes, squids, crabs, shrimps and shellfish.

**BREEDING & CARING FOR YOUNG:** These sharks become adults at around 60–70 centimetres and give birth to live young that feed on other eggs in the uterus. Up to seven young are produced.

**DANGER TO HUMANS:** Angelsharks are not aggressive, but they will bite if they are annoyed.

**WHAT IS THEIR STATUS?** Common, but vulnerable to overfishing.

## Gummy shark *Mustelus antarcticus*

This shark is also known as the Australian smooth hound.

The gummy shark has blunt teeth joined in plates in each jaw. They "gum" their food instead of biting it! The gummy shark does not have scales like a fish but, like all other sharks, has flattened thorns called denticles all over its skin. The skin feels smooth when rubbed towards the tail and like sandpaper when rubbed the other way.

**WHAT DO THEY LOOK LIKE?** The gummy shark is a small to medium-sized, slender-bodied shark with two equal-sized triangular dorsal fins. It has a grey/brown back with many small white spots, and a pale belly.

**SIZE:** Gummy sharks may reach a length of 1.7 metres.

**WHERE DO THEY LIVE?** They are found throughout southern Australia from Geraldton to Port Stevens, down to depths of about 80 metres.

**TEETH & FEEDING HABITS:** They feed on squids, crabs, shrimps and some small fishes.

**BREEDING & CARING FOR YOUNG:** Females become adult at five years and give birth to an average of 14 pups every two years. Females are pregnant for about 12 months. They usually live in small schools that are all the same size and sex. They make long journeys from eastern Tasmania across to Western Australia.

**WHAT IS THEIR STATUS?** Every year, 5000 of these harmless sharks were caught by commercial fishers. This means there are now fewer numbers of these sharks.

# Tawny shark *Nebrius ferrugineus*

Tawny sharks live on coral reefs in tropical waters.

These large sharks are sometimes called sleeper sharks because of their habit of resting or sleeping in caves and crevices during the day. Tawny sharks are blunt-headed with very small eyes, and have a very long upper tail fin. There are two small nose feelers, or barbels, under the chin and the last two gill slits are very close together.

**WHAT DO THEY LOOK LIKE?** They are grey-brown on the back and paler beneath.

**DORSAL FINS:** Tawny sharks have two equal-sized, large dorsal fins well back towards the tail.

**SIZE:** Some tawny sharks grow to a maximum length of about 3.2 metres, but most are smaller than this.

**WHERE DO THEY LIVE?** Tawny sharks live on coral reefs in tropical waters throughout the Indian and west Pacific Oceans. They are common around northern Australia from North West Cape to Rockhampton. They live in shallow water around reefs and are not often found deeper than 50 metres. They may be seen in groups of several individuals lying very close together.

**TEETH & FEEDING HABITS:** They have many closely-grouped, rasp-like teeth in each jaw that are used for gripping and crushing prey. Tawny sharks are mostly active at night, feeding on many different types of prey, including crabs, shrimps, octopuses, squids, sea urchins and small fishes. Prey is sucked into the jaws with a powerful suction made by strong throat muscles. These sharks can reverse this suction to blast out jets of water if they are captured.

**BREEDING & CARING FOR YOUNG:** Tawny sharks become adults at a length of about 2.5 metres and the 2–8 young live in egg cases while they develop. The females keep these egg cases in the uterus until the pups hatch at a length of about 40 centimetres. They are then released as live young. Tawny sharks appear to have a limited home range and are often found resting in the same place during the day. Marine biologists know very little about their behaviour and long-term movements.

**DANGER TO HUMANS:** These sharks are mostly harmless and unafraid of divers. Divers can approach them closely when they are resting during the day, and even touch them. They have been known to bite if they are mishandled or annoyed.

Tawny sharks swim with a very undulating motion.

# Greynurse shark *Carcharias taurus*

The greynurse shark once had a bad reputation as a dangerous man-eater. Its frightening, long, sharp teeth were probably responsible for this image problem. Many of these sharks were killed during the 1950s and 1960s before people realised that they did not usually attack humans. The Australian greynurse shark is also known elsewhere in the world as the sand tiger shark.

**WHAT DO THEY LOOK LIKE?** This shark has a bronze-coloured back and pale belly. There is often a sharp change from dark to pale on the sides. Young greynurse sharks have many darker spots on the tail and back that fade as they grow bigger. These spots are sometimes still visible in large adults.

**DORSAL FINS:** The greynurse shark is a large, bulky shark with a short, pointed snout and a long upper tail fin. The two dorsal fins are about the same size. These medium-sized fins are close together on the back part of the body.

Greynurse sharks have a fearsome set of teeth!

**SIZE:** Maximum length is about 3.2 metres.

**WHERE DO THEY LIVE?** Greynurse sharks live in warm temperate and subtropical waters worldwide. They are only found in two places in Australia — in South-West Western Australia and from southern Queensland to southern New South Wales. They live near the bottom, usually around rock or coral reefs. They are found from close inshore out to the edge of the continental shelf to a depth of 200 metres.

**TEETH & FEEDING HABITS:** There are long, dagger-like teeth in both jaws that can be seen even when the mouth is closed. They feed mainly at night on bony fishes and small sharks and rays. They often hunt and feed in groups with several sharks working together to herd and catch prey.

**BREEDING & CARING FOR YOUNG:** Greynurse sharks become adults when they are about 8 years old and are about 2.2 metres long. After mating, the females produce 16–24 eggs in each womb. Once the young sharks reach 10 centimetres in length, they begin to swim and hunt inside the mother's womb. They eat the other eggs and baby sharks until only a single young shark is left in each womb. The two surviving young feed on unfertilised eggs produced by the mother and are born after a pregnancy of 12 months when they are about 1 metre long. Females breed every second year, so they only give birth to about 10 pups in a lifetime.

**MIGRATION & BEHAVIOUR:** Groups are often found together during the day, swimming slowly along sand-bottom gutters in 15–40 metres of water. Only about 25 resting sites are known for the east coast greynurse sharks and they are now protected. Greynurse sharks make regular migrations of up to 800 kilometres between resting and breeding sites.

Greynurse sharks may reach 3.2 metres in length.

**DANGER TO HUMANS:** Although greynurse sharks have never made an unprovoked attack, they do become excited when fish are speared or caught on a line and should always be treated with care.

**WHAT IS THEIR STATUS?** Rare, protected in Australia.

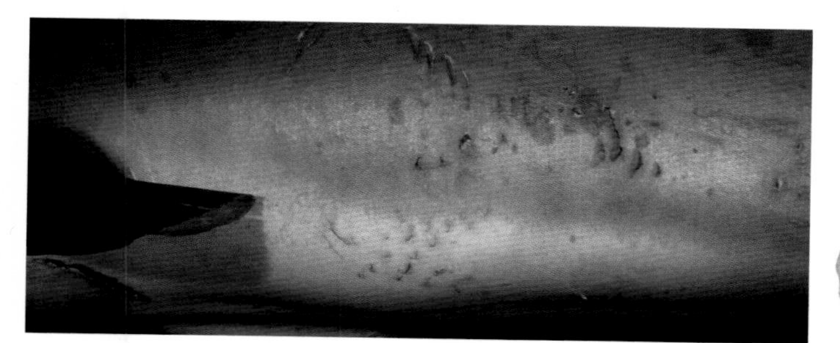

Male greynurse sharks bite females during courting.

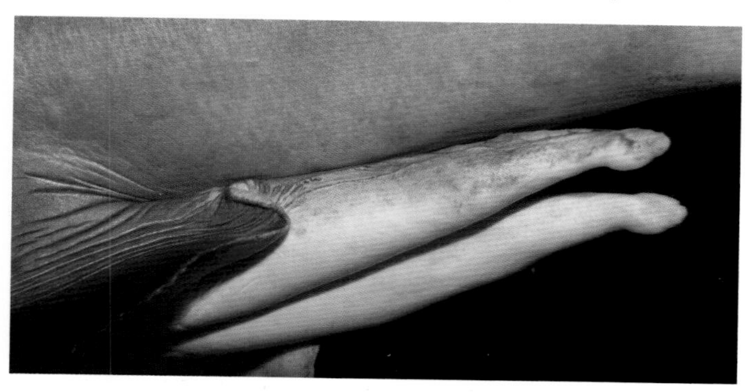

Claspers of a male greynurse shark, used for mating.

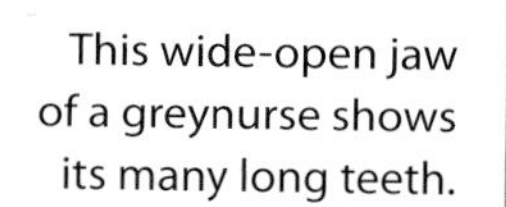

This wide-open jaw of a greynurse shows its many long teeth.

# **Whale shark** *Rhincodon typus*

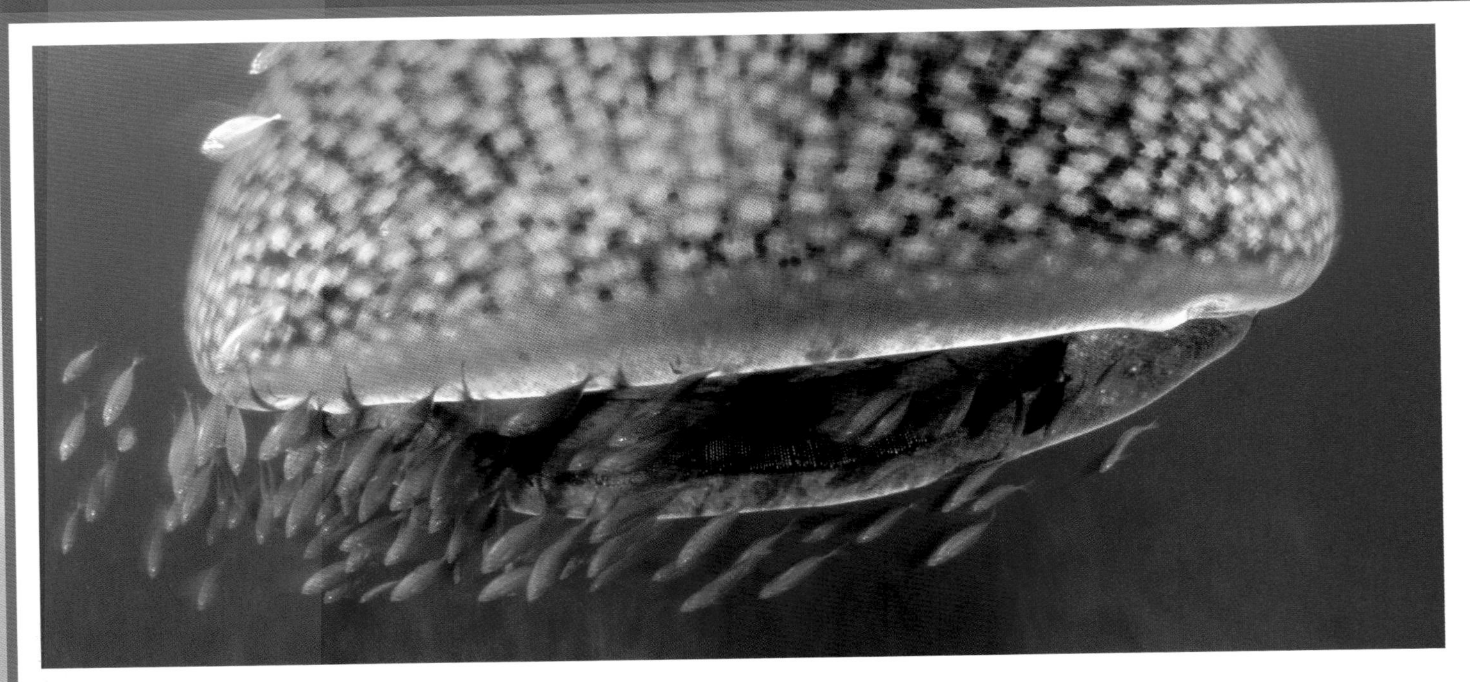

A whale shark feeds on a school of fish.

The whale shark is the world's largest fish. The tail alone of a large whale shark is higher than a man! It has a wide, blunt head and its mouth is at the front of the head, rather than underneath it like most other sharks. Tiny eyes sit on each side of the front of the head. A whale shark's gill slits are very long. The tail is very high and powerful. It might be huge, but the whale shark is a harmless filter-feeder, living on *plankton* animals and tiny fishes.

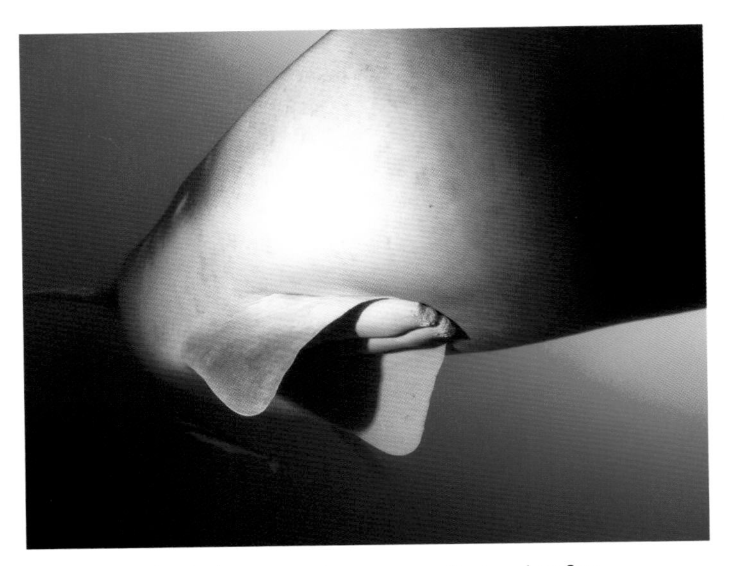

Male whale sharks have two claspers near their pelvic fins.

**WHAT DO THEY LOOK LIKE?** Whale sharks are dark brown on the back and fins, with a checkerboard pattern of white spots and stripes. The belly is white.

**DORSAL FINS:** The large, rounded first dorsal fin sits well back on the body. The second dorsal fin is just in front of the tail and about half as high as the first. The first dorsal and tail fins stick out of the water when whale sharks swim on the surface.

**SIZE:** Most whale sharks are 4–12 metres long, but they may grow as large as 18 metres!

**WHERE DO THEY LIVE?** Whale sharks are found throughout the world in tropical seas. They may be seen all around northern Australia from Perth to Bass Strait. These sharks usually live near the surface in clear water far offshore. Sometimes, they may come close to the coast to feed around reefs and small islands.

**TEETH & FEEDING HABITS:** Whale sharks have lots of tiny pointed teeth. There are about 300 rows of teeth in each jaw. These sharks suck in huge amounts of plankton animals that are then filtered out of the

There is a small eye on each side at the front of the shark's head.

The whale shark is the world's largest fish. They have huge gill slits.

water by a net across the inside of the gill slits. They can also suck in schools of fish and even manage to catch small tuna and squids.

**BREEDING & CARING FOR YOUNG:** Because they are so large and usually live far out in the ocean, we do not know much about the lifestyle of whale sharks. They probably do not become adults until they are about 6 metres long. Females lay their young in large egg cases but it is thought that they are kept inside the female's womb until the pups hatch. These huge sharks are only about 50 centimetres long when they are born!

**MIGRATION & BEHAVIOUR:** Whale sharks migrate long distances to find rich supplies of food. They often gather together in good feeding places. For some unknown reason they make short visits to the surface when they are feeding and this is when they are normally spotted. It is not known how long these sharks live or how many there are in the world.

**DANGER TO HUMANS:** Whale sharks are gentle giants and completely harmless to humans.

**WHAT IS THEIR STATUS?** Whale sharks are rare. Before the mid 1980s only 350 confirmed sightings had been made around the world. Whale sharks were listed in the *Environment Protection and Biodiversity Conservation Act 1999* as vulnerable and migratory.

A whale shark has a unique pattern, just like a human's fingerprint.

# Shortfin mako shark *Isurus oxyrinchus*

The shortfin mako looks similar to its relative, the white shark.

The shortfin mako is the cheetah of the shark world. It can swim very fast and often makes spectacular leaps out of the water. It is a close relative of the white shark and has a similar shaped body. Shortfin makos have a sharp, pointed snout, a crescent-shaped tail, large black eyes and very large gill slits. Like the white shark, the shortfin mako shark can keep its body temperature higher than the water in which it lives.

**WHAT DO THEY LOOK LIKE?** Small makos are bright blue on the back, silver-blue on the sides and white on the belly. Large sharks become more grey-blue in colour. The biggest shortfin makos may look like white sharks.

**DORSAL FINS:** The first dorsal fin of the mako shark is high with a rounded tip. The second dorsal fin is tiny and sits back close to the tail.

**SIZE:** Shortfin makos grow to about 4 metres long.

**WHERE DO THEY LIVE?** These sharks are found worldwide in temperate and tropical seas. They live all around Australia except in the shallow northern waters of the Gulf of Carpentaria and Torres Strait. Shortfin makos live near the surface in offshore waters and are not often seen on the coast. They prefer water that is warmer than 16 degrees Celsius and move closer to the Equator during winter months.

**TEETH & FEEDING HABITS:** This shark has long, sharp teeth in both jaws. Like the greynurse shark, a shortfin mako's teeth can still be seen when its mouth is closed. A shortfin mako feeds only on fast-swimming, small *pelagic* fishes. Its dagger-like teeth are ideal for gripping small slippery fishes while they are swallowed.

**BREEDING & CARING FOR YOUNG:** Males reach adulthood at a length of about 2 metres. As with most species of sharks, the females are bigger and do not become adults until they measure about 2.8 metres. Females produce 4–16 young that are fed in the womb on a supply of unfertilised eggs. Young makos are born measuring about 60–70 centimetres long.

The shortfin mako gets its name from the Maori word *mako*, which means shark.

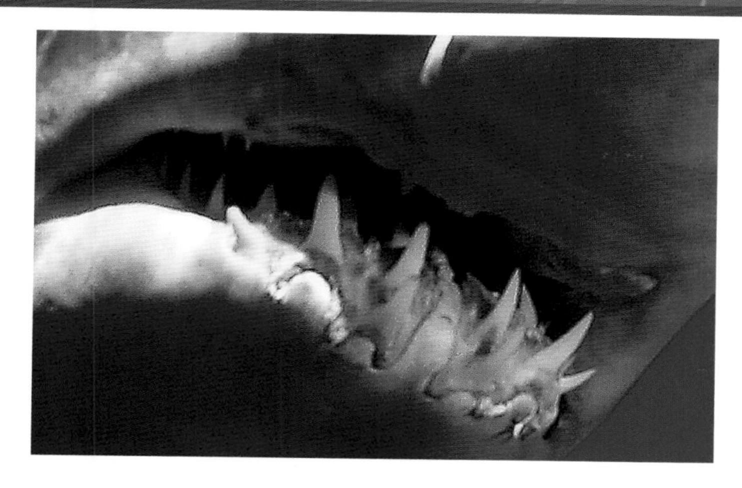

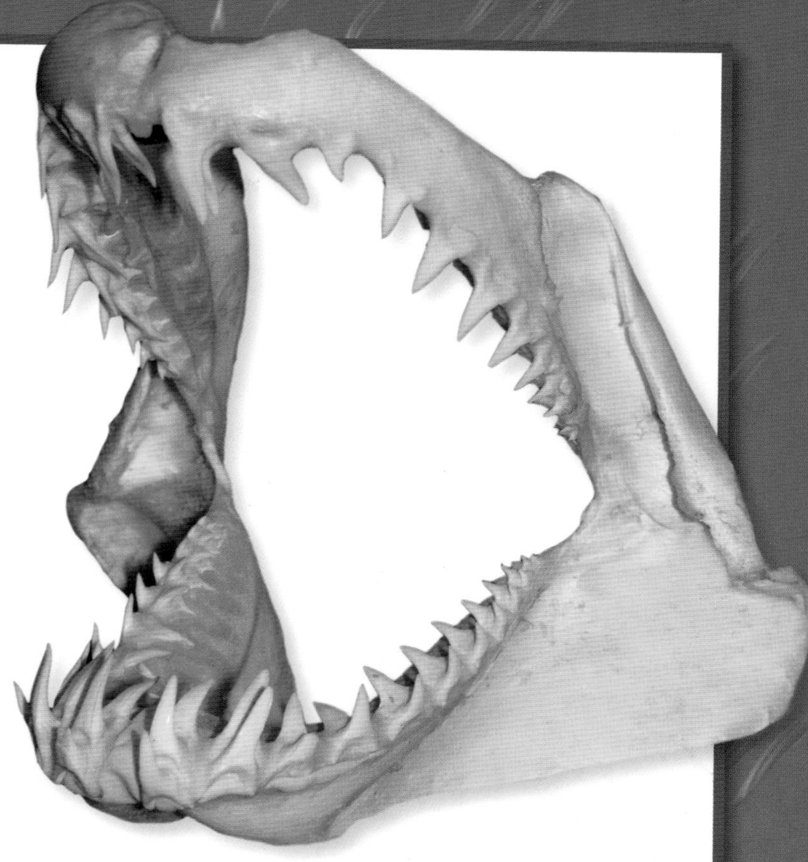

**MIGRATION & BEHAVIOUR:** Like other oceanic sharks, shortfin makos make long migrations. It is not known how they navigate during these journeys across the oceans. There are no landmarks or signposts in the sea, so the sharks probably detect ocean currents or the Earth's magnetic field to discover where they are!

**DANGER TO HUMANS:** Mako sharks only eat small fishes and do not normally attack humans. There are a few reports of these sharks biting swimmers, but these attacks have not been fatal.

*Above and left:* Mako teeth are designed for gripping slippery fishes.

The shortfin mako shark's large black eye gives it a cruel-looking stare, but it is actually quite harmless to humans.

# Elephantfish *Callorhinchus milii*

The elephantfish has a unique look.

The elephantfish is a relative of sharks and rays in a group known as ghost sharks. It is named for the strange blunt *proboscis* that grows from the end of its nose, which looks like the trunk of an elephant. The head is large, with a *tapering* body, large pectoral fins and a long pointed tail fin. Its skin is smooth and has none of the thorn-like denticles that are found on most sharks.

**WHAT DO THEY LOOK LIKE?** Elephantfish are silver-white in colour with dark blotches on the body and fins.

**DORSAL FINS:** The first dorsal fin is high and triangular with a sharp spine in front. This spine is mildly poisonous and makes the elephantfish hard for bigger sharks to eat.

**SIZE:** These fish reach a length of about 1.2 metres.

**WHERE DO THEY LIVE?** Elephantfish are found off southern Australia between Esperance and Sydney. They are abundant south of Bass Strait, where they are trawled commercially and sold as "white fish". The same species is also common around New Zealand. They live on sand or mud bottoms over the continental shelf down to a depth of about 200 metres.

**TEETH & FEEDING HABITS:** The elephantfish has blunt tooth plates designed to crush its food. It feeds on shellfish and sea urchins that it digs from the bottom with its tough, trunk-like nose. When stirred-up silt makes it difficult to see things underwater, the elephantfish uses the sensory organs on its trunk and around its mouth and head to help find food.

**BREEDING & CARING FOR YOUNG:** Elephantfish become adults at a length of about 70 centimetres. They migrate into inshore bays in spring to breed. The females lay several large flask-shaped egg cases that are about 25 centimetres long and have broad side wings. The young hatch after about 8 months.

**DANGER TO HUMANS:** Elephantfish are not a threat to humans. However, if accidentally caught, elephantfish should be handled with care because of their sharp and poisonous fin spine.

**WHAT IS THEIR STATUS?** Common and secure at present. Elephantfish are sometimes caught by commercial fishers and are most abundant off the east coast of New Zealand's South Island.

## Zebra shark *Stegostoma fasciatum*

An adult zebra shark is sometimes called a leopard shark.

Despite its name, the zebra shark is only striped like a zebra when it is very young. Adults are spotted like leopards and are often called "leopard sharks" in Australia. This shark has a stout body and a long, straight tail the same length as its body. Two hard ridges run down the upper body from behind the eyes. Its eyes are tiny and hard to see among the spots.

**WHAT DO THEY LOOK LIKE?** Zebra sharks are yellow or brown with small dark brown spots on the body and fins. Juveniles, or young zebra sharks, are dark brown with vertical yellow stripes.

**DORSAL FINS:** Both dorsal fins are longer than they are high and are set well back on the body.

**SIZE:** Zebra sharks usually reach a length of about 2.5 metres, half of which is tail.

**WHERE DO THEY LIVE?** On shallow reefs around northern Australia between Carnarvon and Sydney.

**TEETH & FEEDING HABITS:** They feed mainly on shellfish, as well as shrimps and small fishes.

**BREEDING & CARING FOR YOUNG:** After mating, females lay a few large eggs that are attached to the bottom by bunches of hair-like strands or "tendrils". The young hatch at a length of about 20 centimetres.

**DANGER TO HUMANS:** These sharks are considered harmless. They may be approached when they are resting on the bottom and can often be touched.

**WHAT IS THEIR STATUS?** Common throughout the Indian and Pacific Oceans.

## Epaulette shark *Hemiscyllium ocellatum*

Epaulette sharks live among coral reefs in shallow water.

The epaulette shark is named for the large black spots on its shoulders, which look a bit like the "epaulettes" on military uniforms. They are probably the most abundant of the coral reef sharks but they rest quietly on the bottom and are difficult to see. These sharks are small and very slender, with small fins and a long, straight tail. Both dorsal fins are long and low on the back half of the body.

**WHAT DO THEY LOOK LIKE?** Epaulette sharks are yellow to pale brown patterned with small dark spots on the body and fins and two large, white-edged black spots on the body behind each gill slit.

**SIZE:** These sharks reach a maximum length of 105 centimetres.

**WHERE DO THEY LIVE?** Epaulette sharks are common hiding among corals in shallow water around reefs in northern Australia between Shark Bay and Moreton Bay. They are also found in waters off Papua New Guinea.

**TEETH & FEEDING HABITS:** Tiny sharp teeth are set in plates in each jaw. These sharks feed on bottom living animals such as shellfish, crabs and sea urchins.

**BREEDING & CARING FOR YOUNG:** Females lay about 50 eggs each year. These oval eggs are about 10 centimetres long and hatch in just four months. Newly hatched young are 15 centimetres long and only grow about 3 centimetres each year.

**DANGER TO HUMANS:** This small shark is harmless.

## Rusty carpetshark *Parascyllium ferrugineum*

These sharks live on the bottom in depths of 40–60 metres.

Many people think that all sharks are large and fierce, but, in fact, more than half of the shark species that live in Australia are less than 1 metre long when they reach their full size! The rusty carpetshark is one of these small sharks. It has a slender body and a straight tail. There is a small feeler, or barbel, next to each nostril under the chin.

**WHAT DO THEY LOOK LIKE?** The rusty carpetshark is grey-brown with six darker brown "saddles" across the back. There are scattered dark brown spots on the body and fins.

**DORSAL FINS:** Two dorsal fins of similar size are set back toward the tail.

**SIZE:** Rusty carpetsharks only reach a maximum length of 80 centimetres.

**WHERE DO THEY LIVE?** These little sharks are only found off southern Australia between Albany in Western Australia and eastern Victoria.

**TEETH & FEEDING HABITS:** A rusty carpetshark's teeth are slender and triangular. They probably feed on small, bottom-living shellfish and crabs.

**BREEDING & CARING FOR YOUNG:** These sharks lay eggs and the young hatch at a length of about 17 centimetres.

**DANGER TO HUMANS:** Very little is known of the habits of the rusty carpetshark but its small size and tiny teeth make it completely harmless to humans.

## Collar carpetshark *Parascyllium collare*

The wide black band or "collar" gives this shark its name.

Like all sharks, the collar carpetshark uses its sensory organs to tell what is happening in its environment. It has a good sense of smell and hearing. It can also detect the weak electrical fields put out by all living animals and uses this electrical sense as the most important way to find its food.

**WHAT DO THEY LOOK LIKE?** The wide, dark brown collar circling the back of the head gives this shark its name. It has a light yellow to reddish brown body, a paler belly, and large dark spots on the body and fins.

**SIZE:** Collar carpetsharks reach a maximum length of just 90 centimetres.

**WHERE DO THEY LIVE?** These sharks live on the bottom, usually near rocky reefs and are common at depths of 60–130 metres. They are sometimes seen in water as shallow as 20 metres.

**TEETH & FEEDING HABITS:** The mouth of a collar carpetshark is small with many small sharp teeth in each jaw. These sharks probably feed on bottom-living shellfish, crabs and sea urchins.

**BREEDING & CARING FOR YOUNG:** Females lay eggs that are left on the bottom to develop for several months before they hatch.

**DANGER TO HUMANS:** This small shark is completely harmless to humans.

# Varied carpetshark *Parascyllium variolatum*

The varied carpetshark looks a bit more like an eel than a shark.

The varied carpetshark has a small head, tiny teeth, and a long slender body more like an eel than a shark. It spends most of its time hiding on the bottom, rather than swimming around hunting prey. This small shark has a mottled, carpet-like colour pattern that *camouflages* its body outline and makes it hard to see when it is resting on the bottom.

**WHAT DO THEY LOOK LIKE?** This shark is grey-brown with a mottled pattern of darker blotches and white spots. There is a large black spot on the tip of all the fins. Behind the head is a wide, dark brown collar covered in small white spots. There are two triangular, equal-sized dorsal fins on the back half of the body.

**SIZE:** This small shark only grows to 90 centimetres.

**WHERE DO THEY LIVE?** They live around the Victorian coast but are also found around northern Tasmania and across southern Australia to just north of Perth. Varied carpetsharks have not been found in southern Tasmania.

They usually live on rocky reefs and seagrass beds, in depths from 5 metres down to about 180 metres. Adult sharks often shelter in cracks and crevices on the bottom, but young sharks may hide completely under rocks during the day.

**TEETH & FEEDING HABITS:** This shark has a small mouth with many small, sharp-pointed teeth in each jaw. Not much is known of the habits of this shark but scientists think they feed on crabs and shrimps, octopuses and small bottom-living fishes. They probably hunt mainly at night.

**BREEDING & CARING FOR YOUNG:** Varied carpetsharks lay egg cases that are then attached to the ocean bottom. The young sharks develop inside these cases until they are ready to hatch. The mothers do not guard their eggs and the young sharks must look after themselves as soon as they are born.

**DANGER TO HUMANS:** These small sharks are not normally aggressive towards humans. Like most animals, they may bite when annoyed but they have small teeth and would not cause much damage.

**WHAT IS THEIR STATUS?** Varied carpetsharks are common, especially in Victoria.

The varied carpetshark spends much of its time on the ocean floor. The shark's blotchy appearance helps it to camouflage.

## Prickly dogfish *Oxynotus bruniensis*

The prickly dogfish's belly is the same colour as its back.

Most dogfish sharks are small but they are long-lived. Some species live more than 70 years. They get the "prickly" part of their common name because their skin is very rough. There are over 40 different species of dogfish sharks found in Australia. Because they usually live in deep water, marine biologists know little about them.

**WHAT DO THEY LOOK LIKE?** The prickly dogfish is brown or dark grey. It has a small head and a large, humped body with a flattened belly. This shark has no anal fin and has a paddle-shaped tail.

**DORSAL FINS:** Both dorsal fins look like high, pointed sails, each with a sharp, mildly poisonous spine in front.

**SIZE:** These sharks reach a length of 72 centimetres.

**WHERE DO THEY LIVE?** They are only found off southern Australia between Esperance and Newcastle and around the coast of New Zealand. They live on the continental slope and are usually found near the bottom in depths from 350–650 metres.

**TEETH & FEEDING HABITS:** The upper teeth are sharp and dagger-like but the lower teeth form a serrated cutting edge. Prickly dogfish eat small, bottom-living fishes as well as crabs and shrimps.

**BREEDING & CARING FOR YOUNG:** These sharks give birth to live young that probably grow slowly.

**DANGER TO HUMANS:** These small sharks are harmless apart from the risk of fin spine injuries.

## Draughtboard shark *Cephaloscyllium laticeps*

This shark is *endemic*, which means it only occurs in Australia.

This shark is sometimes called the "swell shark" because it can blow itself up by swallowing water. This trick makes the draughtboard shark look bigger than it really is so that larger sharks may decide it is too big to eat. The draughtboard shark is a large catshark with broad pectoral fins and a long, notched tail.

**WHAT DO THEY LOOK LIKE?** This shark has mottled brown blotches and irregular darker spots that help camouflage it on the bottom. It has a wide head and large mouth. Both dorsal fins are set well back near the tail. The first is twice as high as the second.

**SIZE:** These sharks are usually less than a metre long but may reach 1.5 metres.

**WHERE DO THEY LIVE?** This shark lives only around southern Australia between Esperance and Jervis Bay. It may be spotted in shallow water but can live in water as deep as 220 metres.

**TEETH & FEEDING HABITS:** These sharks catch crabs, shrimps, squids and small fishes. They often raid crayfish pots, eating the bait as well as the trapped crayfish.

**BREEDING & CARING FOR YOUNG:** Females lay cream-coloured eggs that are about 13 centimetres long. Newly hatched young measure 14 centimetres.

**DANGER TO HUMANS:** Draughtboard sharks are usually considered to be harmless.

**WHAT IS THEIR STATUS?** Common.

## Grey spotted catshark *Asymbolus analis*

These catsharks are able to blend in well with their surrounds.

The grey spotted catshark is a very small slender catshark. There are over 100 different catshark species in the world's oceans and over 30 of these are found in Australia. Catsharks are all small, bottom-living sharks that are rarely seen, except by fishermen on trawler boats who sometimes catch them accidentally in fishing nets.

**WHAT DO THEY LOOK LIKE?** These sharks are grey-brown with darker brown saddles across the back. They are covered in tiny white spots and some larger dark spots.

**SIZE:** A fully grown grey spotted catshark is only 60 centimetres long.

**WHERE DO THEY LIVE?** These small sharks only live in shallow waters off the southern New South Wales coast. They are normally seen resting on the sand near reef outcrops during the day, usually in water less than 60 metres deep.

**TEETH & FEEDING HABITS:** They have small, many-pronged teeth and, like other small catsharks, probably hunt more actively at night. They feed on small fishes as well as squids, crabs and prawns.

**BREEDING & CARING FOR YOUNG:** These catsharks lay flexible cases containing eggs. Female sharks attach the cases to the bottom with sticky hairs or "tendrils".

**DANGER TO HUMANS:** Like all catsharks they are considered harmless to humans.

## Gulf catshark *Asymbolus vincenti*

The gulf catshark has a beautiful mottled pattern.

The gulf catshark is another very small species of catshark with a slender body and small fins. These catsharks have a small hole behind the eye called a *spiracle,* which lets water into the gills when the mouth is closed or when resting against the bottom. Active sharks (which are always swimming) keep their mouths slightly open to pass water over their gills. They do not have a spiracle.

**WHAT DO THEY LOOK LIKE?** The gulf catshark has a mottled pale and dark brown colour pattern. It has many small white spots on the back, sides and tail.

**SIZE:** These sharks are less than 60 centimetres long when fully grown.

**WHERE DO THEY LIVE?** Gulf catsharks are common in the Great Australian Bight and live on the bottom in depths from 130–220 metres.

They are sometimes found in shallow water in Bass Strait where they may be seen by divers close to rocky reefs.

**TEETH & FEEDING HABITS:** These sharks have small teeth and probably feed at night on small fishes, crabs and shrimps.

**BREEDING & CARING FOR YOUNG:** Females lay eggs that are attached to the bottom until they hatch.

**DANGER TO HUMANS:** These small catsharks are very shy and are not regarded as harmful to humans.

**WHAT IS THEIR STATUS?** Common and at their most abundant along the Great Australian Bight.

## Banded wobbegong *Orectolobus ornatus*

This species has two equal-sized dorsal fins set well back.

The banded wobbegong is a well-camouflaged ambush predator. A large, flat body and about ten branched skin flaps around the chin help it to disappear into its surroundings. It usually rests on the bottom waiting for prey. When an unwary fish comes close, this shark makes a lightning-quick lunge off the bottom to grab it.

**WHAT DO THEY LOOK LIKE?** Banded wobbegongs have a mottled colour pattern of yellow, and dark brown saddles and spots.

**SIZE:** Most are less than 2 metres long.

**WHERE DO THEY LIVE?** Banded wobbegongs live around southern and eastern Australia but not in Tasmania. They are also found in Papua New Guinea.

**TEETH & FEEDING HABITS:** These sharks use their long, slender teeth and powerful jaws to catch crabs, crayfish, octopuses and fishes.

**BREEDING & CARING FOR YOUNG:** Like many sharks, male banded wobbegongs give the females love bites when courting! Young wobbegongs feed on unfertilised eggs in the womb and are born at a length of about 20 centimetres.

**DANGER TO HUMANS:** These sharks do not normally attack humans, but will bite and hang on if they are annoyed.

**WHAT IS THEIR STATUS?** These sharks are common, especially around inshore rocky and coral reefs down to about 100 metres.

## Spotted wobbegong *Orectolobus maculatus*

Spotted wobbegongs are often seen by divers.

The common name "wobbegong" is an Aboriginal word, and it is obvious why this wobbegong is called the spotted wobbegong. It is a large species, very similar to the banded wobbegong. Both species are found in southern Australia, but the spotted wobbegong has more branched lobes around the chin than the banded species and has a slightly different colour pattern.

**WHAT DO THEY LOOK LIKE?** These sharks are mottled brown and yellow with distinct circles of small white spots.

**SIZE:** These sharks are usually less than 2 metres long. Some may reach up to 3 metres in length.

**WHERE DO THEY LIVE?** Spotted wobbegongs are common around southern Australia from Perth to Moreton Island. They are not found in Tasmania.

**TEETH & FEEDING HABITS:** These sharks have long, fang-like teeth and feed on crabs, crayfish, octopuses and small fishes.

**BREEDING & CARING FOR YOUNG:** Female spotted wobbegongs give birth to large litters of young. As many as thirty-seven 20-centimetre-long pups have been found in pregnant females!

**DANGER TO HUMANS:** Like other large wobbegongs this species will sometimes bite if injured or annoyed.

**WHAT IS THEIR STATUS?** Vulnerable in New South Wales and listed as near-threatened globally. It is thought that reports of sightings in Japan and the South China Sea are probably mistakes.

## Tasselled wobbegong *Eucrossorhinus dasypogon*

The tasselled wobbegong is almost invisible in its reef home.

The tasselled wobbegong has a very broad flattened body, huge pectoral fins and a flat, rounded head. Around its head is a fringe of branched skin lobes, which looks a little like a beard. This shark has a mottled, camouflage colour pattern making it very hard to see as it rests and waits on the bottom among the corals. It catches unsuspecting fish when they swim too close.

**WHAT DO THEY LOOK LIKE?** These sharks are yellowish or reddish brown with darker blotches and a fine net-like pattern of dark brown lines.

**SIZE:** Most of those caught have measured between 1–1.5 metres, although some experts claim these sharks may grow to more than 3.5 metres.

**WHERE DO THEY LIVE?** Tasselled wobbegongs live in shallow water around coral reefs throughout northern Australia and in Papua New Guinea. They are very well camouflaged and hard to see. The tail is usually coiled in a tight curve when the shark rests on the bottom.

**TEETH & FEEDING HABITS:** These wobbegongs have a large mouth and many sharp-pointed teeth. They ambush fishes and squids and might eat quite large fishes on occasion.

**BREEDING & CARING FOR YOUNG:** Live young are born measuring about 20 centimetres long.

**DANGER TO HUMANS:** These sharks are not aggressive but do have a sharp set of teeth.

**WHAT IS THEIR STATUS?** Common in Australia, but they are listed as near-threatened globally.

## Cobbler wobbegong *Sutorectus tentaculatus*

The cobbler wobbegong is found only in southern Australia.

The cobbler wobbegong is easily recognised by two rows of wart-like lumps that run down the middle of its back. Like other wobbegongs, it has a mottled colour pattern, which makes it hard to see when it is resting motionless on the bottom. Sometimes anglers in South-West Western Australia accidentally catch cobbler wobbegongs on fishing lines around rocky reefs.

**WHAT DO THEY LOOK LIKE?** Cobbler wobbegongs are pale brown with darker brown saddles and blotches and many small black spots on the back and fins. There are only a few small skin lobes around the front of the head.

**SIZE:** These small wobbegongs only grow to about 90 centimetres long.

**WHERE DO THEY LIVE?** Cobbler wobbegongs live on inshore rocky reefs around southern Australia from north of Perth to Adelaide.

**TEETH & FEEDING HABITS:** These sharks have sharp teeth which they use to feed on crabs, shrimps and small fishes.

**BREEDING & CARING FOR YOUNG:** Very little is known about the lifestyle of these sharks. Like all wobbegongs, they give birth to live young that are about 20 centimetres long when they are born.

**DANGER TO HUMANS:** Cobbler wobbegongs do not normally attack humans but they have sharp teeth and powerful jaws and should not be handled or annoyed.

**WHAT IS THEIR STATUS?** Common.

## Scalloped hammerhead *Sphyrna lewini*

Hammerhead sharks have an unusual head shape.

Hammerheads are easy to spot! They have flattened extensions on each side of the head and their eyes are on the ends of these head lobes. The scalloped hammerhead gets its name from the distinct notch in the front of the head. The rest of the body looks much like a regular shark. The upper tail fin is long and pointed and the pectoral fins are strangely small for the size of its body.

**WHAT DO THEY LOOK LIKE?** This shark has a grey-brown back, a pale belly and the odd-shaped "hammer" head.

**DORSAL FINS:** The first dorsal fin is very high. The second is much smaller and set back close to the tail.

**SIZE:** Scalloped hammerhead sharks can reach up to 4 metres in length.

**WHERE DO THEY LIVE?** Scalloped hammerheads are worldwide tropical sharks. They live all around northern Australia, from Perth to Sydney. They prefer surface water around coral reefs but can be found down to 250 metres.

**TEETH & FEEDING HABITS:** They have small mouths with sharp, pointy teeth and usually eat fishes and squids.

**BREEDING & CARING FOR YOUNG:** These sharks give birth in spring after a 9–10 month pregnancy. The 12–24 young are about 50 centimetres long at birth.

**DANGER TO HUMANS:** Hammerhead sharks normally ignore divers and swimmers, but if nearby fishes are caught or speared, they may bite humans.

## Smooth hammerhead *Sphyrna zygaena*

The head hammers may help the shark to turn quickly.

The smooth hammerhead is a cold-water relative of the scalloped hammerhead. Unlike its relation, the smooth hammerhead does not have a notch in the middle of its head and has a lower first dorsal fin. From above, its head extensions or "hammers" look large and clumsy, but they are slim and streamlined when viewed from the side.

**WHAT DO THEY LOOK LIKE?** These sharks have a grey-brown back and a pale belly.

**SIZE:** Smooth hammerheads occasionally reach a length of 4 metres but most are less than 3.5 metres.

**WHERE DO THEY LIVE?** They are found in all temperate seas around the world. They live all around southern Australia, from Perth to Coffs Harbour. These sharks usually live in open water over the continental shelf but may be seen around inshore rocky reefs. Young sharks are often seen swimming on the surface near the coast with their fins out of the water.

**TEETH & FEEDING HABITS:** Like scalloped hammerheads, they often gather in large schools in the same resting spots during the day. They spread out at night to hunt, feeding on squids as well as a few fishes and stingrays.

**BREEDING & CARING FOR YOUNG:** Females give birth to 20–50 young after a 10–11 month pregnancy.

**DANGER TO HUMANS:** They are usually harmless to humans but may bite if excited.

# Great hammerhead *Sphyrna mokarran*

The great hammerhead is one of the largest sharks of the group considered the "dangerous" sharks. It rivals the white shark and tiger shark in length but is not as heavy-bodied. Like all species of hammerheads, it has flattened hammer-like head lobes, but the front of its head is almost straight rather than curved backward like the scalloped and smooth hammerhead. The upper tail fin can be as long as a man is high.

**WHAT DO THEY LOOK LIKE?** This shark has a bronze or grey-brown back and pale belly.

**DORSAL FINS:** The first dorsal fin is very tall and narrow with a curved back. A large great hammerhead may have a 1-metre-high dorsal fin. The second dorsal fin is higher than that of other hammerhead species.

This species has a very tall front dorsal fin.

**SIZE:** Great hammerheads can grow to 6 metres although most are less than 4.5 metres.

**WHERE DO THEY LIVE?** These large sharks live throughout the tropical oceans of the world. They are found all around northern Australia from North West Cape to Sydney.

**TEETH & FEEDING HABITS:** Great hammerheads have small mouths for their size, with sharp, pointed teeth tilted toward the outside of the jaw. These sharks feed on bottom-living fishes and squids. They eat stingrays and often hunt them over sand flats. A large hammerhead was once found with 97 stingray spines embedded in its head!

**BREEDING & CARING FOR YOUNG:** Great hammerheads become adults at a length of about 2.2 metres. Females feed the young in the uterus through a placenta, in the same way that mammals do, and give birth to live young. This means these sharks are *viviparous*. They give birth to 6–33 young after an 11-month pregnancy. Newborn great hammerheads are about 65 centimetres long.

**DANGER TO HUMANS:** Great hammerheads have a bad reputation, but in fact they are not very aggressive towards humans. They normally ignore swimmers and divers but may become excited by fishing or spearfishing activity. A small number of people have been bitten by these excited hammerheads but the attacks have not been fatal.

The great hammerhead can live down to a depth of about 80 metres.

# Manta ray *Manta birostris*

Manta rays glide through the water with slow, graceful flaps of their wings.

Manta rays were once called devilfish because they were thought to be dangerous to humans and even to attack and sink small boats! In fact, they are gentle giants and completely harmless. Manta rays have huge sweptback "wings" that are really modified pectoral fins. They have a very wide mouth on the front of the head and long head flaps on each side of the mouth known as *cephalic lobes*. The tail is short and whip-like.

**WHAT DO THEY LOOK LIKE?** Manta rays have black backs with unevenly patterned white and black bellies.

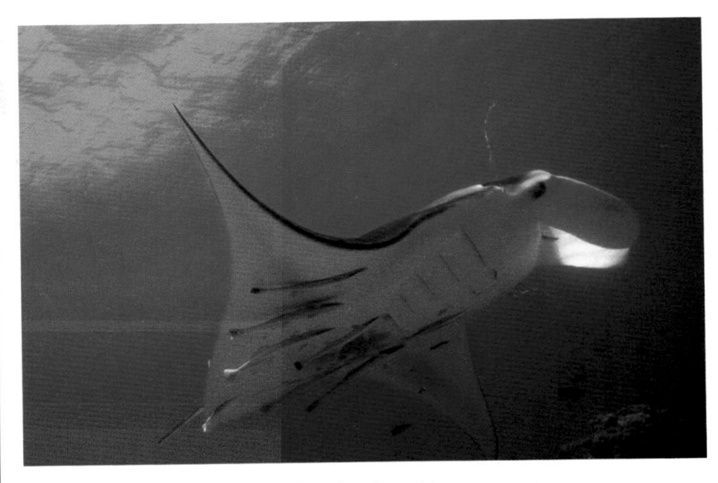

Remoras (or "sucker fish") often hitch a ride on manta rays.

**SIZE:** The manta ray is the largest ray and one of the largest living fishes. It may measure 2–4 metres across and can reach 6 metres across its outspread wings.

**WHERE DO THEY LIVE?** Manta rays are found throughout the tropical oceans of the world. They live all around tropical Australia over the continental shelf, and may venture as far south as Perth or Sydney in summer.

Manta rays fly through the water with slow, graceful flaps of their wings. They are fast swimmers and sometimes make spectacular jumps completely out of the water, falling back with a resounding slap.

**TEETH & FEEDING HABITS:** Manta rays have only a few upper tiny, blunt teeth and feed by swimming along with their mouths wide open to filter plankton animals and small fishes out of the water. The head lobes help to funnel water and food into the mouth. These rays often gather in groups to feed along current lines, where plankton animals are common.

**BREEDING & CARING FOR YOUNG:** No one has ever seen manta rays mating. It is known they give birth to live young, but no other details have been discovered.

**DANGER TO HUMANS:** These giant rays are harmless. They are often curious and will approach divers.

**WHAT IS THEIR STATUS?** Little information is known about numbers of manta rays worldwide.

# Whitespotted eagle ray *Aetobatus narinari*

The whitespotted eagle ray is large, rivalling the manta ray in size. It has wide, pointed "wings" like a manta ray, but has a pointed snout that easily sets it apart from a manta ray. Its tail is long and whip-like and may be over twice the length of its body. Like a stingray, this ray has a spine at the base of the tail, but it is usually much smaller than that of a stingray of similar size.

**WHAT DO THEY LOOK LIKE?** The whitespotted eagle ray has a green-grey back and white underside. It has a beautiful pattern of small white spots on the body and wings.

**SIZE:** If the tail is undamaged they can reach a total length of 9 metres! These rays may be over 3 metres across their outspread wings. However, most of those seen by divers are less than 1.8 metres across.

**WHERE DO THEY LIVE?** Whitespotted eagle rays are found throughout the tropical and warm temperate seas of the world. They live around northern Australia from North West Cape to Brisbane and are often found in shallow water around coral reefs. Unlike most eagle rays and stingrays, they do not spend much time resting on the bottom but are seen actively swimming in open water.

The whitespotted eagle ray is a graceful swimmer.

**TEETH & FEEDING HABITS:** The teeth are joined to form flat plates in each jaw. These are used to crush shellfish such as clams and catseye turban shells, which are the favourite foods of all eagle rays. The rays carefully sort out the meat and leave the crushed shell fragments in a neat pile on the bottom.

**BREEDING & CARING FOR YOUNG:** Whitespotted eagle rays reach adulthood at about five years of age. Females give birth to 2–4 young that are only 25 centimetres across the wings.

**DANGER TO HUMANS:** These large rays are normally harmless to humans but watch out for that spine if they are caught in a net or on a line!

Although similar in size to the manta ray, the whitespotted eagle ray has a pointed snout.

## Sparsely-spotted stingaree *Urolophus paucimaculatus*

The sparsely-spotted stingaree hunts in the sand by feel alone.

The sparsely-spotted stingaree may be one of the smallest members of the ray family, but it can pack a big punch. A long sharp spine with serrated edges on top of the tail can be used as a defensive weapon. The body and head are flattened and hard to see because of the large pectoral fins that give this ray a classic, diamond-shaped body.

**WHAT DO THEY LOOK LIKE?** Sparsely-spotted stingarees are light grey above with a few dark-edged white spots on the back.

**SIZE:** This ray only reaches 45 centimetres in length, including its tail.

**WHERE DO THEY LIVE?** Sparsely-spotted stingarees are found off southern Australia from Perth to northern New South Wales. They live on sandy bottoms in water down to about 150 metres in depth.

**TEETH & FEEDING HABITS:** Like the teeth of most rays, the sparsely-spotted stingray's teeth are fused into bony crushing plates.

A sparsely-spotted ray blows water through its mouth to create holes in the sand and expose shellfish, sea urchins and worms.

**BREEDING & CARING FOR YOUNG:** Like stingrays, these rays give birth to live young that are miniature versions of the adults.

**DANGER TO HUMANS:** If threatened, this ray will strike strongly upward with its poisonous tail spine and may cause a nasty wound.

## Smooth stingray *Dasyatis brevicaudata*

Smooth stingrays are usually seen resting on sand or mud.

Smooth stingrays are the largest of all the stingray species and may weigh as much as four big men put together. The poisonous tail spine on a full-sized smooth stingray is over 40 centimetres long and makes a deadly weapon. The whip-like tail is shorter than the body and covered in small thorns. When trodden on, these rays have accidentally stabbed people, causing death.

**WHAT DO THEY LOOK LIKE?** Smooth stingrays have a black back and a white belly with a thin black edge.

**SIZE:** They may reach 2 metres across the body disc and a length of over 4 metres including the tail.

**WHERE DO THEY LIVE?** They live around southern Australia from Shark Bay to Brisbane, and are also found in New Zealand and South Africa. They are common in shallow water to depths of 100 metres.

**TEETH & FEEDING HABITS:** Bony tooth plates are used to grip and crush shellfish and animals dug from the sand.

**BREEDING & CARING FOR YOUNG:** They may gather in large courting groups, either resting on the bottom or cruising slowly through underwater caves and archways. Each female has several young that are born at a disc width of about 35 centimetres.

**DANGER TO HUMANS:** This ray may arch its tail over its back when approached but is usually curious rather than aggressive. These large rays should always be treated with great care.

**WHAT IS THEIR STATUS?** Common.

# Southern fiddler ray *Trygonorrhina fasciata*

The southern fiddler ray lives in shallow water.

With a bit of imagination, this ray looks a little like a musical fiddle, which is how it got its name. It is part of the shovelnose ray family, with a head and body like a ray but two dorsal fins and a tail like a shark. It does not have the tail spine of a normal ray; instead it has a line of small thorns down the centre of the back.

**WHAT DO THEY LOOK LIKE?** The southern fiddler ray is yellow-brown on the back and white beneath. There is a pattern of bright blue-white lines on the head and back and each line has dark brown edges. There are darker brown blotches on the head and tail and between some of the line markings. Young fiddler rays have darker markings and brighter blue lines than the adults.

**SIZE:** Southern fiddler rays can grow to a length of 1.25 metres but most of those seen are less than 1 metre long.

**WHERE DO THEY LIVE?** Southern fiddler rays live in shallow water between Perth and Bass Strait. They are often found in seagrass estuaries and off sand beaches and may be seen around jetties and wharfs. A very similar looking species called the eastern fiddler ray is found off southern Queensland and the east coast of New South Wales.

**TEETH & FEEDING HABITS:** The teeth are blunt and joined into plates. These rays may scavenge on dead animals and will often enter fish or cray traps. They also hunt for crabs, shrimps and small fishes.

**BREEDING & CARING FOR YOUNG:** Female southern fiddler rays produce golden egg cases. Each case holds 2 or 3 young rays. These egg cases are held in the mother's uterus until the young hatch. The young are born alive.

**DANGER TO HUMANS:** Southern fiddler rays have a small mouth, blunt teeth and no tail spine. They are harmless to humans.

**WHAT IS THEIR STATUS?** These rays only live in waters around Australia but they are common and especially abundant in southern Western Australia.

The southern fiddler ray is one of the more colourful species of shovelnose ray.

# Blotched fantail ray *Taeniura meyeni*

These large rays are almost circular.

The blotched fantail ray is a huge stingray with a disc-width of 1.8 metres. Divers often have heart-stopping moments when they are approached, and sometimes even bumped, by these curious rays. The disc is almost circular and the short tail has a long, fin-like tip. There is usually one long tail spine but some large rays have a second, smaller spine.

**WHAT DO THEY LOOK LIKE?** As its name suggests, the blotched fantail ray has a back patterned with black blotches and a white belly.

**SIZE:** The blotched fantail ray can grow to a length of about 3.3 metres, including the tail.

**WHERE DO THEY LIVE?** The blotched fantail ray is found throughout tropical parts of the Indian and Pacific Oceans. Although it is not well known in Australia, it is probably found in all coral reef areas and inhabits waters from the central coast of Western Australia, Australia's tropical north and down the south-east coast to northern New South Wales and Lord Howe Island.

**TEETH & FEEDING HABITS:** Blotched fantail rays feed on animals dug from the sand including, shellfish, crabs, shrimps and sea urchins.

**DANGER TO HUMANS:** Although they are not aggressive, blotched fantail rays have been known to use their long spines to kill humans. They should not be annoyed or approached too closely.

**WHAT IS THEIR STATUS?** Common.

# Common stingaree *Trygonoptera testacea*

Common stingarees are found in coastal sand and reef areas.

This stingaree is another small stingaree species. It has a short-finned tail and two small sharp tail spines. It swims by passing waves down each side of the disk rather than flapping its "wings". These waves alternate from one side to the other and the ray glides slowly along, hardly disturbing the sand. Common stingarees swim very fast when they are threatened.

**WHAT DO THEY LOOK LIKE?** It has a brown or grey back with pale disc edges and a white belly. There is a single tiny dorsal fin in front of the tail spines.

**SIZE:** A full-grown common stingaree is less than 30 centimetres across the disc and less than 50 centimetres long including the tail.

**WHERE DO THEY LIVE?** They are found in shallow water off southern Queensland and New South Wales. These rays are sometimes caught in offshore waters down to depths of about 60 metres.

**TEETH & FEEDING HABITS:** Joined teeth plates are used to grip and crush food, including shellfish, sea urchins, crabs and worms dug from the sand bottom.

**BREEDING & CARING FOR YOUNG:** Stingarees develop their young inside paired wombs using a placenta. They have 2–4 young that are born after a 3-month pregnancy.

**DANGER TO HUMANS:** Common stingarees are small but, like all rays, should be treated with care.

**WHAT IS THEIR STATUS?** Common.

## Spotted stingaree *Urolophus gigas*

The spotted stingaree has a densely patterned back.

The spotted stingaree has a nearly circular disc. Its back is closely patterned in white-grey spots. It uses the sense organs around its mouth and its excellent sense of smell to help it identify its food — because its eyes are on top of its head, it cannot see what goes into its mouth.

**WHAT DO THEY LOOK LIKE?** Spotted stingarees are covered in densely packed pale spots on a dark brown or black background. The small dorsal fin and paddle-shaped tail are black with white margins.

**SIZE:** This is one of the largest members of the stingaree family and reaches a total length of about 70 centimetres with a disc width of 40 centimetres.

**WHERE DO THEY LIVE?** Spotted stingarees live in shallow water around southern Australia from Albany in Western Australia to Bass Strait.

**TEETH & FEEDING HABITS:** These rays feed on animals that burrow in the sand. They eat shellfish, crabs, worms and sea urchins.

**BREEDING & CARING FOR YOUNG:** Spotted stingarees give birth to 2–4 live young.

**DANGER TO HUMANS:** The tail spine of a spotted ray is only about 8 centimetres long but it can cause a nasty cut if the ray is aggravated or feels threatened. Like all rays, the spotted stingaree should be treated with care and respect.

**WHAT IS THEIR STATUS?** Common.

## Banded stingaree *Urolophus cruciatus*

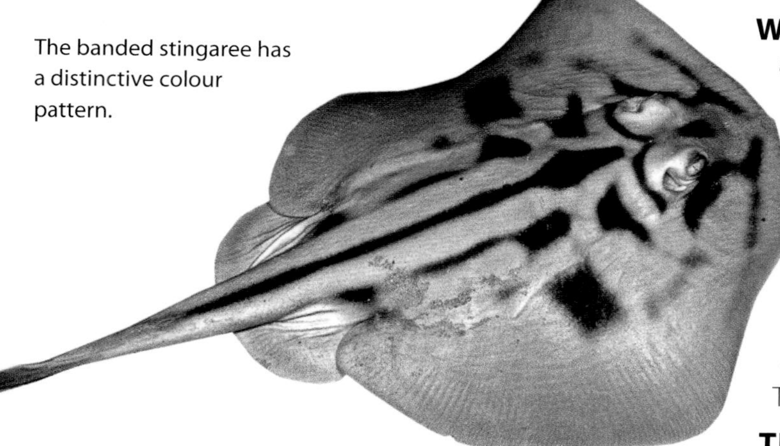

The banded stingaree has a distinctive colour pattern.

The banded stingaree is a small ray with a very short tail and a striking colour pattern. It spends most of its time resting on the ocean floor on sand. To hide from predators, it partly covers itself with sand. It does this by quickly fluttering its wings to raise a cloud of sand that then settles over its body. The banded stingaree then lies still, almost invisible against the sand.

**WHAT DO THEY LOOK LIKE?** Banded stingarees are usually pale grey or yellowish brown with a pattern of dark bars on top of the disc. The belly is white, sometimes with darker blotches.

**SIZE:** These rays grow to about 30 centimetres across the disc and to a total length of 50 centimetres, including the tail.

**WHERE DO THEY LIVE?** Banded stingarees are found on the continental shelf down to depths of about 160 metres through Bass Strait and around Tasmania.

**TEETH & FEEDING HABITS:** These rays feed on small sand-living animals like shellfish and crabs.

**BREEDING & CARING FOR YOUNG:** Male banded stingarees may become adult at a length of only 25 centimetres. Females give birth to a small number of live young.

**DANGER TO HUMANS:** Like most rays, the banded stingaree should to treated carefully to avoid the sharp, tail spine.

**WHAT IS THEIR STATUS?** Common.

## Eastern shovelnose ray *Aptychotrema rostrata*

The eastern shovelnose ray blends easily into the sandy bottom.

The eastern shovelnose ray is often called a guitarfish, because its outline looks just like a guitar. Although it is a ray, the back part of its body, dorsal fins and tail looks a lot like a shark. It does not have a tail spine to protect it from predators but this ray can swim faster than most rays using rapid *undulations*, or rippling movements, of its body and tail.

**WHAT DO THEY LOOK LIKE?** They have a pale grey-brown back, sometimes with a pattern of darker blotches, and a paler, long, pointed snout.

**SIZE:** These rays sometimes grow to a length of 1.2 metres but most are less than 85 centimetres long.

**WHERE DO THEY LIVE?** They are common in estuaries and off sandy beaches on the New South Wales coast.

**TEETH & FEEDING HABITS:** These rays have many small, blunt teeth in each jaw. They are active hunters and scavengers and dig in the sand for shellfish and other sand-burrowing animals, as well as crabs and small fishes.

**BREEDING & CARING FOR YOUNG:** Shovelnose rays often bury themselves in the sand when resting. Females become adults at less than 70 centimetres long and give birth to live young. There are 2–8 young rays in each egg case, which stay inside the mother until they hatch.

**DANGER TO HUMANS:** Eastern shovelnose rays are harmless.

## Giant shovelnose ray *Rhinobatos typus*

The giant shovelnose ray may grow to 2.7 metres.

The giant shovelnose ray is the largest of the shovelnose rays. It has two high, pointed dorsal fins and looks like a shark with a flattened and sharp-pointed head. Sucker fish like to attach themselves to large sharks and rays for a free ride and to eat the scraps from their meals. Large sucker fish are often found attached to giant shovelnose rays.

**WHAT DO THEY LOOK LIKE?** Giant shovelnose rays are grey to greenish-brown on the back with a white belly. The pointed snout is often yellowish with a dark stripe down the middle.

**SIZE:** These giants grow to a length of 2.7 metres.

**WHERE DO THEY LIVE?** They live around tropical Australia and through Indonesia to India. Young are often found in very shallow water in mangroves and off beaches. Adults may live down to depths of 100 metres and are sometimes seen on sand patches around coral reefs.

**TEETH & FEEDING HABITS:** These rays have small, blunt teeth and feed mainly on shellfish dug from the sand. They may sometimes eat small fishes.

**BREEDING & CARING FOR YOUNG:** They give birth to a small number of live young.

**DANGER TO HUMANS:** Although large, these rays have small mouths, blunt teeth, no fin or tail spines and are considered harmless to humans.

**WHAT IS THEIR STATUS?** These rays are near threatened in Australia and globally vulnerable.

# Melbourne skate *Dipturus whitleyi*

A large Melbourne skate weighs 50 kilograms.

About 40 of the world's 200 species of skate live in Australian waters. The Melbourne skate is the largest Australian species. Like all skates, the Melbourne skate has a flattened, ray-like body with a long pointed snout, two small dorsal fins back at the end of a short thin tail, many thorns on its back and tail but no long tail spine.

**WHAT DO THEY LOOK LIKE?** They are pale grey or brown on the back and white underneath with scattered small white spots on the back. Young Melbourne skates have a large dark blotch on each side of the disc.

**SIZE:** These skates grow to a length of 1.7 metres and are 1.2 metres across the disc.

**WHERE DO THEY LIVE?** Melbourne skates are not only found around Melbourne. They live around southern Australia from Albany in Western Australia to just south of Sydney. They live in shallow water and are common close to shore but may also be found in water down to 170 metres deep.

**TEETH & FEEDING HABITS:** Skates have many small teeth. They dig in the sand for crabs, shrimps and small fishes.

**BREEDING & CARING FOR YOUNG:** Most skates lay eggs in horny egg cases that are stuck to the bottom with sticky coiled hairs at each corner.

**DANGER TO HUMANS:** The Melbourne skate's thorny tail may cause grazes and cuts.

# Thornback skate *Dipturus lemprieri*

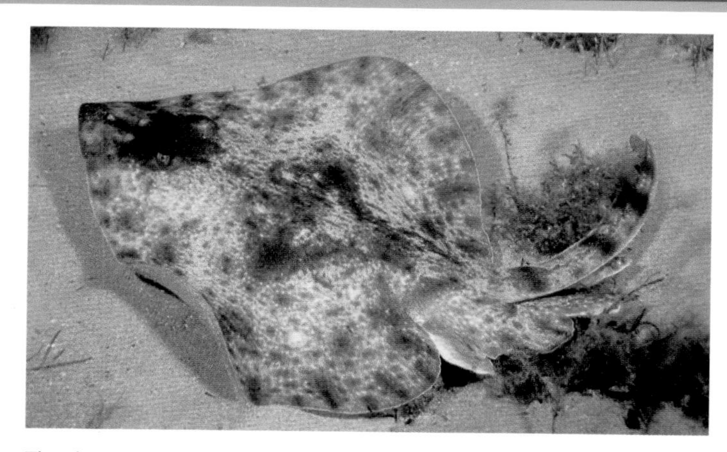

The thornback skate has a blunter snout than most skate species.

Each species of skate has a different pattern of thorns on its skin. The thornback skate has thorns above each eye, down the middle of the back and on the tail. Males have thorns that can be pulled down into the skin in a patch on each wing and may be used during mating. Male skates have long claspers each side of the tail to hold the females when mating.

**WHAT DO THEY LOOK LIKE?** Thornback skates have blunt snouts compared with many skates. They have grey-black or brown backs covered with a pattern of darker blotches and fine, net-like markings.

**SIZE:** These skates grow to 50 centimetres.

**WHERE DO THEY LIVE?** Thornback skates live over sandy bottoms out to depths of 170 metres but are also common in shallow water. They are found through Bass Strait and around Tasmania.

**TEETH & FEEDING HABITS:** Thornback skates feed on crabs, shrimps and other sand-living animals.

**BREEDING & CARING FOR YOUNG:** Little is known about the lifestyle of most of the skates. Male thornbacks become adults at 40 centimetres long. After mating, the females lay a number of eggs that take some months to hatch at about 15 centimetres long.

**DANGER TO HUMANS:** The tail thorns of skates may cause grazes or cuts if they are handled or annoyed. Care should be taken with the thornback skate.

**WHAT IS THEIR STATUS?** Common.

## Shark ray *Rhina ancylostoma*

The shark ray is called a "bowmouth" because of its wide mouth.

The shark ray looks like a shark with a flattened head and very large pectoral fins. It is actually a large ray and may weigh as much as two men. Behind its eyes are two large spiracles. Shark rays are sometimes caught in trawler nets and may damage the rest of the fishermen's catch.

**WHAT DO THEY LOOK LIKE?** The shark ray has two high dorsal fins and a high, shark-like tail. It has a blue-grey back with many large white spots on the body and fins. There is a large black spot at the base of each pectoral fin and dark bands between the eyes. These markings fade as the ray gets older.

**SIZE:** This shark ray grows to a length of 2.7 metres.

**WHERE DO THEY LIVE?** They are found in shallow water throughout the tropical Indian and Pacific Oceans. Shark rays are not often seen in Australia but may be seen around the northern coasts, usually near coral reefs.

**TEETH & FEEDING HABITS:** Many small flattened teeth form plates to crush shellfish and crabs.

**BREEDING & CARING FOR YOUNG:** Not much is known of the lifestyle of the shark ray. It probably gives birth to live young that develop in egg cases in the female, like the shovelnose rays.

**DANGER TO HUMANS:** Shark rays are very large and although they have blunt teeth and no tail spine they should be treated carefully.

## Coffin ray *Hypnos monopterygium*

This ray is also called a "numbfish" because it gives electric shocks.

The coffin ray is an electric ray. It has a powerful biological battery in each side of the body disc that can be switched on by the ray to give a strong electric shock to any animal near it or touching it. Electric rays use electric shocks to stun the animals and fishes they eat and to frighten off predators. The coffin ray is often called a "numbfish" because of the effects of these electric shocks.

**WHAT DO THEY LOOK LIKE?** Coffin rays are many different colours. They may be reddish brown, chocolate brown, grey or pink and often have a few darker or paler spots and blotches. They are white on the belly.

**SIZE:** Most coffin rays are less than 40 centimetres long including the tiny, stumpy tail.

**WHERE DO THEY LIVE?** They live off the coast of New South Wales. They are also found between Adelaide and Broome on the south and west coasts of Australia. Coffin rays mainly live close to the coast but have been caught by trawlers in water as deep as 220 metres.

**TEETH & FEEDING HABITS:** They use their small, three-pronged teeth to eat crabs, worms and small fishes.

**BREEDING & CARING FOR YOUNG:** These rays give birth to 10-centimetre-long live young.

**DANGER TO HUMANS:** The electric shocks given off by the coffin ray are strong enough to throw a person to the ground, but they are not fatal.

## Green sawfish *Pristis zijsron*

The green sawfish has a sharp, saw-like nose half as long as its body.

The green sawfish gets its name from its long flattened "saw" (with its line of long, sharp teeth down each side) that grows from its snout. This dangerous saw is almost half the length of the body. This fish looks like a shark but is actually an altered ray. The green sawfish is the longest of the rays and grows to a very large size.

**WHAT DO THEY LOOK LIKE?** Green sawfish have two high, triangular dorsal fins and a high, shark-like tail. They have a greenish-brown back and a white belly.

**SIZE:** This sawfish grows to more than 7 metres long, including the saw.

**WHERE DO THEY LIVE?** The green sawfish lives in the tropics from South Africa to Australia. It is found throughout northern Australia usually in shallow water and often swims into estuaries and rivers.

**TEETH & FEEDING HABITS:** This sawfish thrashes its saw violently to and fro to stun and cut up fishes, crabs and shrimps. This sawfish uses its small, flattened teeth to pick up the pieces and eat them.

**BREEDING & CARING FOR YOUNG:** Male green sawfish become adults when they are 4.3 metres long. The young develop inside the female's two uteruses until they are born.

**DANGER TO HUMANS:** Green sawfish are not usually aggressive but can be dangerous if trapped.

**WHAT IS THEIR STATUS?** Endangered globally.

## Common sawshark *Pristiophorus cirratus*

The common sawshark has a long bony snout.

Sawsharks look very similar to sawfishes but they are true sharks and not modified rays. The gill slits are on the side of the body, as in sharks, rather than underneath as in sawfishes and other rays. Common sawsharks have long, flattened bony snouts with sharp teeth down each side. There is a long sensory feeler halfway along each side of the saw, like a moustache.

**WHAT DO THEY LOOK LIKE?** The common sawshark has a long shark-like body with two equal-sized dorsal fins. It has a pale yellowish or reddish-brown back with darker bands, blotches and spots. The saw is pinkish with dark bands.

**SIZE:** They grow to 1.4 metres in length.

**WHERE DO THEY LIVE?** These sawsharks live on the bottom in water between depths of 40–300 metres off southern Australia from Perth to Bass Strait.

**TEETH & FEEDING HABITS:** The sawshark uses its sensory feelers to find prey hidden in the sand, then digs it up and kills it with its saw. It feeds on crabs, shrimps, small fishes and other sand-living animals.

**BREEDING & CARING FOR YOUNG:** Common sawsharks become adults at a length of 1 metre. They give birth to live young born at 38 centimetres long.

**DANGER TO HUMANS:** Sawsharks caught in a net or on a line should be handled carefully as the saw could cause some damage to humans.

**WHAT IS THEIR STATUS?** This species is often caught in nets, but it is not considered threatened.

# SHARK CONSERVATION

Shark netting of beaches kills hundreds of large sharks each year.

## WHY ARE SHARKS UNDER THREAT?

For each person killed by a shark there are over 10,000 tonnes of sharks and rays killed by fishermen in Australia. Most of the world's commercial shark fisheries have declining catches because more sharks are being caught than are being born. Female sharks give birth to low numbers of young during their life, and many sharks are not replacing their numbers as fast as humans are killing them. Most recreational fishermen also kill the sharks they catch. In the 1950s and 1960s spearfishermen killed large numbers of sharks in an effort to save people from being attacked. Shark netting off beaches in Queensland and New South Wales is successful at keeping sharks away from humans, but only because it kills sharks that live near beaches faster than they can be replaced. Some sharks are in serious trouble.

Scientists think that nine out of every ten reef sharks on the Great Barrier Reef have been caught over recent years. Most of these sharks have their fins cut off and dried to be sent to Asia to make shark fin soup. Many people now realise that this cannot go on.

## HOW CAN WE HELP?

Restrictions are being put on commercial shark catches to try to protect common shark species. Rare sharks — such as white sharks, basking sharks and greynurse sharks — are now completely protected in Australia and in many other countries. Sharks are some of the largest predators in the oceans; it is important for sharks to survive so the *ecology* of our seas can remain healthy.

Spearfishermen once killed many sharks to protect people from shark attack.

Each year, the dried fins of thousands of sharks are sent to Asia to make soup.

# SHARK ATTACK

## KEEPING IT IN PERSPECTIVE

The thought of being eaten by a shark terrifies most people. Many people won't even go swimming because of the risk of being attacked by a shark. But sharks have much more to worry about from human attack than humans have to worry about sharks. Each year hundreds of thousands of Australians swim or dive in the sea but on average, only one person a year in Australia dies from a shark attack. This compares to one death a year from crocodile attacks, two deaths from lightning strikes, two from bee stings, two from surfboard accidents, nine from diving accidents, 320 from drowning and over 2500 deaths from motor vehicle accidents!

As you can see, it is certainly more risky to get in a car than it is to take a chance in the ocean with sharks.

As the below table shows, even if you were to be attacked by a shark you probably have a better than 50% chance of survival. In many States the percentage of fatal attacks is less than 30%.

**Total shark attacks and fatalities over a 200 year period**
(Data collected by the Australian Shark Attack File).

| Total Attacks | State | Fatal Attacks | Percentage of Fatal Attacks |
|---|---|---|---|
| 244 | NSW | 72 | 29.5% |
| 228 | QLD | 72 | 31.6% |
| 34 | VIC | 7 | 20.5% |
| 50 | SA | 21 | 42% |
| 79 | WA | 13 | 16.5% |
| 12 | NT | 3 | 25% |
| 21 | TAS | 5 | 28.9% |

The above table shows the total number of shark attacks, both provoked and unprovoked, in all Australian States.

## SHARK SIZE IN PERSPECTIVE

Like mammals, which range in size from tiny shrews and mice to huge elephants and gigantic blue whales, sharks come in a huge range of sizes. Whale sharks, the gentle giants of the shark world, grow to over 12 metres in length. Basking sharks, harmless filter-feeders of cold waters, reach a length of 10 metres. The largest of the dangerous sharks, the white, tiger and great hammerhead sharks, all grow to about 6 metres long. On the other hand, over half the Australian shark species do not grow to more than 1 metre in length. The smallest shark is the tiny pygmy shark that lives in the open ocean and is only 20 centimetres long when fully grown.

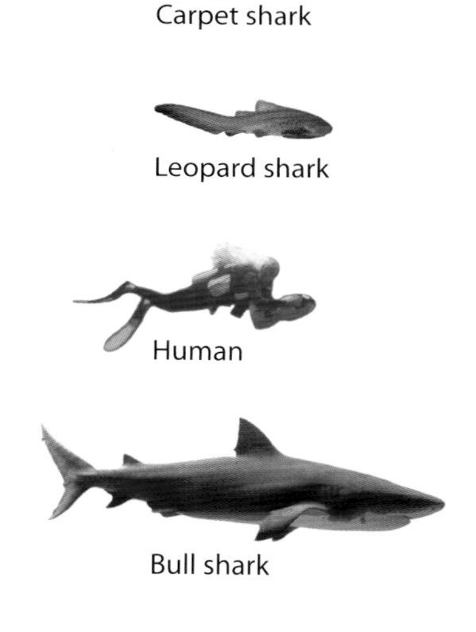

Sharks come in a huge range of sizes.

# GLOSSARY

**Barbels** Slender, fleshy feelers around the mouths of some types of fish.

**Breed** To produce young that can also have their own babies.

**Camouflage** Colours that help an animal blend into its background.

**Carnivore** An animal that eats meat and other animals.

**Cartilage** Flexible connective tissue.

**Cephalic lobes** Long head flaps on each side of the mouth of manta rays.

**Claspers** A tube-like part of the pelvic fin of male sharks and rays by which sperm are introduced to the female.

**Continental shelf** Part of the continent that is under shallow seas.

**Denticles** Tooth-like scales of sharks, rays and skates.

**DNA** (abbreviation for **deoxyribonucleic acid**) Material that transmits inheritable information in genes from parent to offspring.

**Dorsal fins** The fins on the back of fishes and sharks.

**Ecology** The part of biology that deals with the relations between organisms and their environment.

**Embryo** A young animal in an early stage of development, before birth or hatching.

**Endemic** An animal or plant that can only be found in one particular location.

**Estuary** The part of a river that is affected by ocean tides.

**Evolved** To change through descending generations.

**Extinct** A species no longer living on Earth.

**Flange** A projecting rim, edge or flap.

**Habitat** The place in nature where a certain kind of animal lives and breeds or where a plant grows.

**Inhabit** To live in a certain place.

**Intra-uterine cannibalism** When the larger and stronger embryo eats the weaker embryo.

**Invertebrates** An animal that does not have a backbone.

**Lobe** A piece of flesh or skin that sticks out.

**Mate** When animals mate, the male transfers special cells called sperm to the females eggs, which causes young animals to develop. "Mate" also means a partner.

**Migration** To move from one place to another at the same time of year.

**Nocturnal** Active during the night.

**Pectoral fins** The fins on the side of fishes and sharks.

**Pelagic** Of the open ocean or animals living there.

**Pelvic fins** The fins that are on the underside of fishes and sharks.

**Placenta** A structure that connects the circulatory system of an unborn baby to its mother.

**Plankton** Plants and animals that drift in the sea, usually microscopic.

**Predator** An animal that preys on others for food.

**Prey** Animals that are hunted and eaten by other animals.

**Proboscis** A long, tubular snout or feeding organ.

**Reproduce** Animals that have young are said to "reproduce". The natural way for a species to continue.

**Serrated** Having a saw-like edge.

**Spawn** To release eggs or sperm.

**Species** A group of animals that shares the same features and can breed together to produce fertile young.

**Spiracle** An opening in the head of sharks and rays through which water is drawn and passed over gills.

**Tapering** To become gradually slender towards one end.

**Undulation** A rippling motion.

**Uterus** A female's womb.

**Viviparous** Giving birth to live young rather than eggs.

**Vulnerable** Easily hurt or close to extinction.

**Widespread** Occurring over a large area.

**Yolk sac** An initial source of nutrients inside the egg of some species of sharks.

# INDEX